Victor Moritz Goldschmidt: Father of Modern Geochemistry

Brian Mason

National Museum of Natural History, Smithsonian Institution, Washington, D.C. 20560

Special Publication No. 4

THE GEOCHEMICAL SOCIETY

Published by The Geochemical Society
Division of Sciences, Mathematics and Engineering
Trinity University
715 Stadium Drive
San Antonio, Texas 78212 U.S.A.

Library of Congress Catalogue Number 91-078244

ISBN 0–941809–03–X

Printed in The United States of America
by Lancaster Press, Inc.

CONTENTS

Preface

THIS BIOGRAPHY SHOULD have been written in 1950, when most of Goldschmidt's friends and colleagues were still alive. In 1971 Professor Denis Shaw of McMaster University (Hamilton, Ontario, Canada) began collecting material for such a biography by writing to surviving friends and colleagues for reminiscences, but no publication ensued. The initial stimulus for the preparation of the present biography was provided by Professor Charles Sclar of Lehigh University (Bethlehem, Pennsylvania). Early in 1986, being aware of the impending centenary of Goldschmidt's birth in 1988, he proposed to The Geochemical Society that this event should be suitably celebrated. At its annual meeting in May, 1986, the council of the society "nominated C. Sclar (Lehigh University) to form a committee with representatives of other interested societies . . . to bring a proposal to council at its 1986 Fall Meeting for the commemoration of V. M. Goldschmidt's birth in 1888." Sclar formed a committee with himself as chairman and the following members: M. L. Crawford (Mineralogical Association of Canada representative), M. Fleischer, D. M. Henderson, M. J. Holdaway (Mineralogical Society of America representative), G. Kullerud (Meteoritical Society representative), and B. Mason (Geochemical Society representative). The committee met at Lehigh University in December, 1986, and made a number of recommendations, the principal ones being the holding of a Goldschmidt Centennial Symposium in May, 1988, and the preparation of a Goldschmidt biography. Professor G. Kullerud, a Norwegian geologist then on the faculty of Purdue University (Lafayette, Indiana), informed the committee that Aslak Kvalheim, Goldschmidt's long-time associate and his successor as Director of the Norwegian Raw Materials Laboratory, had planned such a biography but had suffered a stroke and was unable to continue. He had offered Kullerud access to his material if he would write the biography. The committee recommended that Kullerud accept this offer. Financial support was forthcoming from

Norwegian sources, and Professor Kullerud spent some time during the summers of 1987, 1988, and 1989 organizing the Goldschmidt Archives in the Geological Survey of Norway in Trondheim. He had begun to write an introductory chapter before his untimely death in October, 1989.

In March, 1990, Dr. Knut Heier, Director of the Geological Survey of Norway, approached me to accept this responsibility. I spent two weeks in July, 1990, and two months in April–June, 1991, perusing these archives, comprising 144 file boxes. Goldschmidt corresponded widely, and copies of his letters, memoranda, and some unpublished manuscripts had been carefully preserved. Unfortunately, Goldschmidt was an only child, never married, and left no close relatives, so many details of his personal life are unavailable.

Why the title "Father of Modern Geochemistry"? The concept of an autonomous discipline dealing with the chemistry of the Earth is an old one, the term "geochemistry" having been introduced in 1838 by the Swiss chemist Schönbein (discoverer of ozone). Throughout the nineteenth century, geochemical data were mainly the byproduct of general geological and mineralogical investigations and comprised more and better analyses of the various units—minerals, rocks, natural waters, and gases—making up the accessible parts of the Earth. The results of these investigations were admirably summarized by F. W. Clarke in 1908, in his monograph *The Data of Geochemistry* (U.S. Geological Survey Bulletin 330). Clarke was Chief Chemist of the U.S. Geological Survey from 1883 to 1925, and was responsible for a vast and ever-growing output of analyses of rocks, minerals, and ores collected by the field staff or submitted for examination. His monograph passed through five editions in less than twenty years, the last in 1924. In many respects this marked the end of an era in geochemistry. During the preceding hundred years geochemical research was largely synonymous with the analysis of those parts of the Earth accessible to visual inspection and chemical assay. From the nature of things it could be little more; interpretative geochemistry, the creation of a philosophy out of the mass of factual information, had to wait upon the development of the fundamental sciences, physics and chemistry. Fundamental advancements in these sciences were made in the early years of this century, such as the discovery and exploration of radioactivity, the Rutherford-Bohr atomic structure, the discovery of X-ray diffraction and its application to chemical analysis and crystal structure. It is the mark of Goldschmidt's genius that he seized upon these discoveries. His insight and intuition, his ability to plan and expedite extensive research programs, and not least his recruitment and inspiration of devoted research associates, revolutionized geochemistry. Thanks largely to his work and stimulus, geochemistry has developed from a somewhat incoherent collection of factual data to a philosophical science based on the geochemical cycle, in which the individual elements play their part according to established principles.

When I began this biography I was inspired by the following statement by M. F. Perutz, in his preface to *Is Science Necessary* (E. P. Dutton, New York, 1989):

> "There is little benefit in following scientists' daily grind but much in tracing the unique combinations of theoretical knowledge and manual skills, the web of personal encounters and accidental observations, the experience, temperament, moods and clashes that go into the making of discoveries, even though the crucial leap of the mind is often impenetrable. There is also something to be said for finding out why others, seemingly just as able, were too blind to grasp what Nature tried to tell them."

Goldschmidt epitomized the characteristics described by Perutz. In writing this book I have tried to interpret this for the possible readership, which as I see it may comprise three groups: geologists, chemists, and geochemists; historians of science; and the layman interested in science but not necessarily conversant with chemistry and geology. The task is a difficult one; what may be obvious to the scientist is far from obvious to the layman, so I have attempted to provide enough scientific background to inform the latter, hopefully without boring other prospective readers.

Most of the documents in the Goldschmidt Archives are written in German and Norwegian. In translating these into English, I have tried to translate as literally as possible, but into idiomatic English. Unless otherwise indicated, quoted letters and documents are preserved in the Goldschmidt Archives in Trondheim.

Careful readers may note some inconsistencies in the ionic radii used in the text. This results from Goldschmidt using 1.33Å for the radius of O^{2-}, whereas the current value is 1.40Å.

One of the intriguing features of Goldschmidt's scientific career is his change from field to laboratory research around 1917. This may not have been solely due to the setting up of the Raw Materials Laboratory at that time. Goldschmidt's health after 1917 did not permit strenuous field work, although he delighted in taking colleagues, students, and visitors into the field and demonstrating the source of many of his ideas.

There is no denying a dark side to Goldschmidt's personality. As the correspondence quoted in this book will show, he inspired lifelong devotion in most of his students and research associates. With his colleagues in the University of Oslo, however, the story was different. He was unduly sensitive to questions of priority and credit, and he had a suspicious nature—in his office at the museum all the cabinets had individual padlocks, because of probable imaginary fears that his research materials might be stolen. His feud with his old student, Professor Barth, seriously affected faculty relationships at the University of Oslo, and saddened many who, as did I, liked and admired both men.

Brian Mason
Washington, D.C.
November, 1991

Acknowledgements

I AM GREATLY INDEBTED to the following colleagues, who read the complete manuscript and whose comments have been extremely helpful: Robin Brett (U.S. Geological Survey); Bjørn Bølviken, Knut Heier (Geological Survey of Norway); Johannes Dons (Geological Museum, Oslo); Alfred Levinson (University of Calgary); Charles Sclar (Lehigh University); Denis Shaw (McMaster University); Cecil Schneer (University of New Hampshire); and Ellis Yochelson (Smithsonian Institution). Hans Wedepohl (University of Göttingen) provided interesting material and photographs from Goldschmidt's Göttingen days. Gerhard Regnell and Frans–Erik Wickman (University of Lund) provided copies of the Hadding-Goldschmidt correspondence, Carl Francis (Harvard University) copies of the Palache–Goldschmidt correspondence, and Kalervo Rankama and Martti Lehtinen (Helsinki University) copies of the Eskola–Goldschmidt correspondence. Professor Ivan Rosenquist, Dr. Leif Gjessing, and Dr. Nils Houge (Oslo) gave me their personal reminiscences of Goldschmidt. Gustaf Arrhenius (University of California, San Diego) was especially helpful on the relationship between his father–in–law, George de Hevesy, and Goldschmidt. I thank Mrs. Ruth Kullerud for the papers of her late husband, Gunnar Kullerud, relating to his proposed biography. Mrs. Harm Buning–Thomassen provided much useful information on the close association of her father, Lars Thomassen, and Goldschmidt in the 1919–1929 period. I thank my brother Alan for his help in compiling the glossary.

Alfred Levinson and Charles Sclar assembled most of the photographs reproduced in this book. Many of the photographs and the color plate appeared in *Applied Geochemistry* (Vol. 3, No. 4, 1988) and are reproduced with the kind permission of Dr. Brian Hitchon, Executive Editor, and Pergamon Press.

Financial support for this work has been provided by Norges Teknisk–Naturvitenskaplige Forskningsråd (Royal Norwegian Council for Scientific and Industrial Research) and the Scholarly Studies Fund of the Smithsonian Institution, and is gratefully acknowledged.

1 / Prologue

I FIRST RAN ACROSS the name V. M. Goldschmidt in 1937, when I was a fourth–year student at Canterbury University College in Christchurch, New Zealand. Browsing through the latest journals, I saw his Hugo Müller Lecture (March 17, 1937) to the Chemical Society of London, entitled "The Principles of Distribution of the Chemical Elements in Minerals and Rocks." At the time I was doing some research on the electrochemistry of dilute solutions, and this lecture opened up an entrancing new field of surpassing interest, one that had never been mentioned in my courses in chemistry and geology. I checked *Chemical Abstracts* for his other publications, only to find that they were mostly in German journals unavailable in New Zealand. The only one I located was entitled "Rare Elements in Coal Ashes," published in *Industrial and Engineering Chemistry*, 1935, and this was also fascinating.

In March 1939 I was working in the New Zealand outback as an exploration geologist for a subsidiary of the Shell Oil Company when I received the offer of a Graduate Fellowship from the University of New Zealand—£200[1] a year for two years of overseas study. Traditionally these scholarships were taken up at universities in England, but I broke tradition by asking to take it up in Norway. I wrote to Goldschmidt explaining my circumstances and asked if I might come to his institute and do research in geochemistry. I received a delightful reply (unfortunately lost during the war), saying that he had never had a student from the Antipodes, that he would be glad to have me, and he assured me that in Oslo I could "live like a king on £200 a year." There was no request for transcripts or any other evidence of my academic qualifications—university life was simpler in those days! It took some months to terminate my contract with the oil company and I finally left New Zealand for Norway in November 1939; World War II had begun in September but was now in its "phony phase."

[1] See *Notes*, starting on page 117, for this and all subsequent footnotes.

I sailed to San Francisco and spent the month of December leisurely crossing the U.S. by Greyhound bus. During this time I met Adolf Pabst,[2] who had worked with Goldschmidt in 1928–29 and gave me much good advice. I also met Tom. Barth[3] (Professor of Petrology at the University of Oslo, who was at that time at the Geophysical Laboratory in Washington), from whom I learned of the serious rift between him and Goldschmidt, a touchy subject.

I arrived in Oslo one evening early in January 1940 and next morning went to the Geological Museum,[4] for which Goldschmidt was director and where he had his research facilities, mainly emission and X–ray spectrographs. The secretary greeted me warmly, said the professor was expecting me, and ushered me into the inner sanctum. Goldschmidt was an impressive figure, heavily built, and with a thick black moustache. At that time he was 52 years old. After a brief chit–chat, the great man announced "Today you must have dinner with me. At home I have a steak as big as a taxicab!" He took me for a tour of the research facilities, introduced me to his assistants and fellow research students, allotted me a room, and cautioned me against smoking on the same floor as the spectrographs. He had already selected my research theme, the geochemistry of tellurium.

Shortly after noon he called for a taxi and we drove to his house in Holmenkollen, a suburb of Oslo with a magnificent view of the city and fjord. The famous international ski jump was just behind his house. Goldschmidt was a bachelor, and had a cook-housekeeper, Miss Marie Brendingen.[5] We went into the kitchen, and there lay the steak—if not as big as a taxicab, it at least covered most of the table. "Mr Mason, you cannot tell me from what beast this steak has been cut." "Yes, Professor, it is a whale steak." I could wish that all my quizzes had been as easy! As a child I lived in Port Chalmers, a small town in southern New Zealand, where my father was superintendent of the dock. This town was the first and last call of the Norwegian whaling fleet then operating in the Ross Sea, and whale steak became part of our diet when the whalers returned at the end of the summer season. So I was fortunate in starting on a good footing with my professor. We had an excellent dinner, starting with lobster (he urged me to eat lots of lobster so as to support the Norwegian fishermen, whose markets had been disrupted by the war) and followed by whale steak and onions, washed down with a good red wine, finishing with dessert, coffee, and brandy. He tried to get me to smoke one of his big fat cigars, but I prudently declined. I was in a distinctly elevated mood when he called for a taxi to take me to the university hostel where he had arranged my accommodation. As I was about to leave, he said "Of course you read German." What a dilemma—should I confess my inability, thereby exposing myself as unqualified for my research responsibilities, or should I bluff it out? Actually I said nothing, whereupon he pressed a large bundle of his reprints on me, saying "Just browse through these before you come to the museum tomorrow.'' When I got to my room I started on one, got completely lost in the first paragraph, a single sentence of several hundred words, and gave up in despair, seeing myself exposed as an unworthy ignoramus the next morning (actually, he never brought up the subject, and over the next few weeks, with the aid of a dictionary and the help of my fellow students, I gradually acquired a reasonable facility in scientific German).

In 1940 essentially nothing was known of the abundance and distribution of tellurium (Te) in rocks and minerals, nor of methods for concentrating this element, so I had an open field ahead of me. We started work around 8 a.m., and around 11 a.m. the geochemical group would gather together to eat sandwiches and drink coffee. During this break Goldschmidt usually appeared and discussed our individual work, asking questions and making suggestions. He proposed measuring tellurium concentrations on two lines in the far ultraviolet of the emission spectrum, which proved to be sensitive down to about 1 ppm (parts per million). I made standards by coprecipitating known amounts of Te with PbS (lead sulfide). Goldschmidt then had me measure Te in olivine ($(Mg,Fe)_2SiO_4$), nickel-iron, and troilite (FeS), all separated from meteorites, which showed (as expected from the position of Te in the Periodic Table) that this element is concentrated in the troilite and to a lesser extent in the metal, and rejected by the olivine. I had started analysing for Te in terrestrial minerals when the German invasion of April 9, 1940, abruptly terminated this research. It received a brief mention in Goldschmidt's book *Geochemistry,* which was published in 1954.

I left Oslo on the morning of the invasion with the university lecturer in English, who had taken his car out of winter storage the previous day; we were a couple of hours ahead of the German army. I had no opportunity to speak to Goldschmidt, but my Norwegian friends told me he spent the day burying his valuables in his garden. After some interesting adventures I reached Stockholm, where I was destined to remain until August 1943 (although I did not anticipate it at the time). I enrolled at the University of Stockholm, but was unable to continue my tellurium research. Instead, I began crystal structure research in the Chemistry Department with Professor Westgren and mineralogical research in the Mineralogical Institute with Professor Quensel. From time to time I had word of developments in Oslo. Conditions there deteriorated severely in 1942 when the Germans established a so-called ''national government'' and installed Vidkun Quisling as ''ministerpresident,'' a figurehead Führer (and Gulbrand Lunde,[6] a former Goldschmidt research associate, as Minister of Propaganda). Anti–Jewish regulations were promulgated (the Jews in Norway were a tiny minority, fewer than 2000 in the whole country). In October 1942 most Jews, including Goldschmidt, were arrested and put in the Berg concentration camp, on the outskirts of

Tønsberg. Eventually they were deported to camps in Poland, where most of them died.

One evening in December 1942 I was working in the Chemistry Department on interminable calculations having to do with the crystal structure of the mineral braunite, Mn_7SiO_{12} (an experience that weaned me from crystal structure analysis for ever). The phone rang and it was Professor Westgren, who invited me to his house, which was quite nearby, to meet an unspecified visitor. Imaging my surprise to walk into the living room and find that the visitor was none other than Goldschmidt, much thinner than I remembered him, but in excellent spirits. He had been rescued by the Norwegian underground and transported to Sweden by way of a remote border crossing. He was particularly proud of the fine pair of boots the underground had provided; they were stolen from German army stores.

He remained in Stockholm until March 1943, when the British authorities arranged to fly him to England. I recall one day he came into the Mineralogical Institute in high glee, waving a postcard and saying "Her har jeg ett brevkort fra Guds bestefar im Zürich" (Here I have a postcard from God's grandfather in Zürich). "God's grandfather" was his reference to Professor Paul Niggli, an eminent mineralogist with a markedly pontifical manner.

I got to England in the fall of 1943, by which time he had settled at the Macaulay Institute for Soil Research in Aberdeen, Scotland. In December I visited him; we spent a pleasant evening in his apartment in front of a good coal fire. He drew my attention to the ash, which was a bright yellow color; he had procured the coal from the Hartley seam in the Durham coalfield, in whose ash he had shown, some years earlier, the presence of 1–2% germanium (GeO_2 has a bright yellow color).

That was the last time we met. I returned to New Zealand in 1944, and took up a position as lecturer in the Geology Department of Canterbury University College. We corresponded from time to time. Shortly after the end of the war in Europe I received a long letter from him in England, to which I replied (September 19, 1945):

> "My dear Professor, Thank you very much for your letter of July 3, which arrived here this morning. It was indeed a pleasure to hear from you again. I was extremely sorry to hear of your long illness during the last winter, and I do hope that you have completely recovered and are feeling fit and well once more. I am looking forward to seeing a long series of scientific papers from your pen for many years to come. It was gratifying to learn that your instruments, library, laboratory notes, and manuscripts had been rescued from your flat in Oslo, and had remained in safe keeping throughout the occupation. I will be interested to hear how you find things in Oslo on your return. It is indeed unfortunate that you have had so much opposition from a section of your university

colleagues. I hope you will get a satisfactory settlement of the whole matter. Please give my very best wishes and regards to my old friends and associates at the Museum, especially Mr. Kvalheim. A short time ago I received two parcels from Norway via Stockholm, one containing my university diplomas, and the other containing my thesis for the Master of Science degree of the University of New Zealand. To those who so carefully looked after these papers throughout the war years and then forwarded them to me my best thanks are due.

My heartiest congratulations to you on your election to an honorary fellowship of the Chemical Society of London. I am very glad to see that British scientific institutions have recognised the epoch–making advances you have achieved in geochemistry and honoured you accordingly. I am hoping that the publication of your book on geochemistry will not be long delayed. It will fill a very serious gap in the literature of science, and will make known the content and implications of geochemistry to a wide circle of scientific men who at present have very little understanding of the subject. So much of your work is inaccessible to British workers, as it is published in periodicals which are not normally available in our libraries. I endeavour to teach a certain amount of the fundamentals of geochemistry to my students, but find myself handicapped by lack of a suitable work of reference, especially as practically all my scientific papers are still in Stockholm."

In his last letter, dated December 1, 1945, he wrote:

"My return to Oslo is somewhat delayed because the dockers strike has caused more than 2 months delay in my large scale silicosis experiments.[7] Anyhow both in Norway and here my work is in excellent progress. I hope to be able to send you some reprints concerning this matter in a near future.

My Oslo laboratories are getting on excellently, and I am getting a lot of new spectrographic data for my book from there.

Meanwhile there has been a "show down" at the University, resulting in complete victory. The rector and the senate sent me a letter much welcoming my return and so did, of course, the Ministry of Commerce. I think the matter has been rather awkward for certain persons, who have good reason to regret their foolishness. I am very glad that you got your certificates [my university diplomas]—which for 5 years were hidden behind my reprints, in the anteroom to the spectrography department. When we put them at that place we very much wondered, in 1940, what the world might be like, when we could dispatch them to you.

My health has been improving, so that I could give a lecture on X-rays and geochemistry before the Royal Institution, two weeks ago, to which I was invited in connection with the 50–years anniversary of the discovery of X–rays. At the Tøyen Museum now all assistants and students have returned, some after many adventures, but fortunately all very much alive.

I am very sorry that you could not continue your tellurium work, as that element is the most unknown in all geochemistry, and we are very much in need of reliable data. Couldn't you follow up your discovery, that tellurium is so common in galenite [lead sulfide] ores, and make some quantitative chemical determinations in galenites and also in lead-antimony sulphosalts which probably are still richer in tellurium? Such an investigation would be most attractive and also has practical importance, especially in gold–mining continents.

My book has been somewhat delayed because of the immense material available. Some ideas concerning the trends of development will be seen from the article in *Soil Science.*

I can very well understand that you are very much attracted by the petrology of metamorphic rocks. I was bound by the same spell for very many years also. And most of my results is still unpublished because I fell for the charms of geochemistry.

And now a lot of my time has been devoted to the problem of preventing silicosis in foundries, and thus keeping disablement and death from that trade, by means of applied geochemistry. Anyhow, there is lots and lots to do for much more mineralogists than exist, or are likely to exist in any reasonable future.

I shall be glad whenever you recommend to one of your research students to make use of the facilities of our laboratories, which are as good as they have ever been. And we expect a number of British scientists there, in the years to come. And I think with much gratitude and satisfaction on the generous hospitality I have had in Britain for nearly three years now.

I always shall be very glad to hear from you and I hope you and your wife are enjoying life on the antipodic point about of the Northern countries.

When I come to Oslo, I shall try to send you copies of some of my old work on the metamorphic rocks of Norway, which I think may be more easily available than my geochemical publications" (original English).

From this letter it can be seen that he was looking forward with optimism and enthusiasm to returning to Norway and continuing his work there. Sad to say, when he returned to Oslo in June 1946 he was a very sick man, and he died shortly thereafter, on March 20, 1947.

2 / *Early Years: 1888–1906*

VICTOR MORITZ GOLDSCHMIDT was born in Zürich on January 27, 1888, the only child of Heinrich Jacob and Amelie Goldschmidt. He was named Victor after Victor Meyer,[1] a famous chemist who was a colleague of his father, and Moritz after his maternal grandfather Moritz Koehne. Heinrich Jacob Goldschmidt was born in Prague on December 4, 1857, the son of a factory owner whose family had lived in Bohemia for many generations. The family can be traced back to the year 1600 and it was pure Jewish. Almost all the family was highly educated; most of them, at least in the early days, were rabbis, but in later generations there were numerous professors, physicians, judges, lawyers, and some military officers. Heinrich Jacob had been intended for a career in industry, but his life interest was chemistry, and after attending a commerce school he entered the University of Prague and received his doctor's degree in chemistry in 1881. In the autumn of 1881 he was appointed docent in chemistry at the Eidgenossische Polytechnikum in Zürich (now the Eidgenossische Technische Hochschule), and in 1885 he was promoted to professor. He married Amelie Koehne in Berlin on March 27, 1887. She was born in Danzig (then in Prussia; now Gdansk, Poland) on December 30, 1864, the daughter of a lumber merchant. The family was evidently well-to-do and she had received an excellent education, augmented by extensive travels in Europe with her parents. Although her home was in Danzig, the family evidently spent some of their winters in Berlin; her diaries mention the theater and concert seasons of 1883–84 and 1884–85 in that city, and it was probably here that she first became acquainted with her future husband.

Heinrich Goldschmidt moved to the University of Amsterdam in 1893, in order to work with J. H. van't Hoff,[2] perhaps the foremost physical chemist of that time. He remained there until 1896, when on the recommendation of Victor Meyer, he was appointed extraordinary professor of chemistry at the University of Heidelberg. Victor

Moritz probably began his schooling at the age of six in Amsterdam, and then attended a private school in Heidelberg. His friend and biographer, Paul Rosbaud,[3] tells the following story from that time, which sheds some light on Goldschmidt's personality. His mother was described as a dignified lady who taught young Victor always to be polite and also tell the truth, however disagreeable. After a few days at his new school the headmaster, a big man with a black beard and of alarming appearance, asked the young man how he liked the school, and ended by saying: "I hope we will become good friends. Do you already like me?" Aware of his mother's admonition, after a moment's pause he replied: "Nicht so besonders, Herr Göckel" (Not particularly, Mr. Göckel).

Heinrich Goldschmidt did not remain long in Heidelberg. In 1900 the chair in chemistry at the University of Oslo became vacant. He applied, and in November he was appointed by the unanimous decision of the Akademiske Collegium. Heidelberg tried to retain him by the offer of an ordinary professorship in physical chemistry, but Heinrich considered himself committed to the position in Norway. He arrived in Oslo with his wife and 13–year old son in April 1901.

Norway in 1901, although a small country on the outer margin of Europe, was in the midst of an intellectual and political renaissance. Some historical background may be in order. At the end of the Viking Age, around 1000 A.D., Norway was an independent kingdom with colonies in Iceland, the Faroes, and Greenland. King Harald Hardraade founded the city of Oslo in 1050; he claimed the English throne in 1066, and was defeated and killed at the battle of Stamford Bridge (near York), shortly before the battle of Hastings. In 1397 Norway was united with Sweden and Denmark under a single king. Sweden broke away in 1523, but Norway remained united with Denmark, and was ultimately reduced to a province. Danish became the official language, and much of the administration was appointed from Copenhagen. After a great fire in 1624 the ancient capital of Oslo was rebuilt and named Christiania in honor of the reigning Christian IV of Denmark; this was changed to Kristiania in 1877, and the historical name Oslo was reinstated from January 1, 1925.

During the Napoleonic wars Denmark was allied with France, much to the detriment of Norway, whose commercial interests were closely linked with England. After the defeat of Napoleon at Leipzig in 1813, Sweden attacked Denmark, and at the Peace of Kiel (January, 1814) Denmark ceded to Sweden its rights to Norway (but not its rights to the old Norwegian colonies of Iceland, the Faroes, and Greenland).

A considerable faction in Norway was opposed to union with Sweden. The governor of Norway, the Danish Prince Christian Frederick (later King Christian VIII of Denmark, 1839–48), called a constitutional convention, which on May 17, 1814, declared independence and elected Christian Frederick to the Norwegian throne. Sweden threatened war; Christian Frederick resigned and Norway became the junior partner in a united kingdom of Sweden and Norway. The year 1814 is remembered in Norway for its national day, celebrated annually on May 17.

The union with Sweden was not popular in Norway, and by 1901 independence had long become a political objective. This was stimulated by a strong sense of nationhood, exemplified by the achievements of native sons, such as the explorers Fridjof Nansen, Otto Sverdrup, and Roald Amundsen, the dramatists Bjørnstjerne Bjørnson and Henrik Ibsen, the painter Edvard Munch, and the composer Edvard Grieg. So although small (the population of Oslo in 1901 was only 227,000) and somewhat remote, Norway was a stimulating environment for the Goldschmidt family. The university had some professors of international renown, including Waldemar Christopher Brøgger[4] in geology.

The university was established by Frederik VI, king of Denmark and Norway, by an edict dated September 2, 1811, and he named it Universitas Regia Fredericiana (in Norwegian Det Koniglike Frederiks Universitet; the name was changed in 1939 to Universitetet i Oslo). Frederik VI endowed the university with a considerable area of land at Tøyen, then on the eastern outskirts of Oslo. Teaching began in 1813 with five professors and 17 students. The university began its operations in rented quarters. The first permanent buildings were erected in downtown Oslo, the foundation stone being laid in 1841 on the thirtieth anniversary of its establishment, although the buildings were not completed until 1852. Three science museums, for zoology, botany, and geology, were built in the Tøyen area during 1911–17. By 1930 the university had outgrown its downtown area, and plans were made for a new campus on a farm at Blindern, in the northern part of the city. Most of the science faculties moved to Blindern during the 1930's, and the remaining institutions moved there after World War II. The university now (1990) has some 25,000 students and 3500 employees. The operations of the university are governed by an elected Rector (the first was Brøgger in 1907) and the Akademiske Collegium.

Victor Moritz evidently learned Norwegian rapidly and could soon read and write his new language with ease. His spoken Norwegian was grammatically correct and very eloquent, but throughout his life he retained a pronounced accent which informed the listener that Norwegian was not his mother tongue. On arrival in Oslo he was enrolled in the Vestheim Middle School (approximately equivalent to junior high school), and in June 1903 he entered a private Gymnasium (high school) from which he graduated in June 1905 with a grade point average of "meget tilfredsstillende" (very satisfactory). He passed written examinations in Norwegian, German, Mathematics, and Latin, and oral examinations in Norwegian, German, English, French, Latin, History, Geography, Natural Sciences, and Mathematics. He was exempted from participation in the gymnastics program required by Norwegian schools, evidently because his physical abilities were not on the same level as those of most of his fellow students. Although he seemed robust, his health was never good, and this probably tended to make him somewhat of a loner. Late in his life he commented: "Sometimes I regret that I never went out and drank beer with my fellow students."

The year 1905 was an eventful one, both for Norway and for the Goldschmidt family. The union with Sweden was peacefully dissolved, and Prince Carl of Denmark was elected king, assuming the title of Haakon VII (Haakon VI, who ruled from 1355 to 1380, was the last king of independent Norway). The Goldschmidt family became Norwegian citizens, and in September he matriculated at the University of Oslo, where he studied geology, mineralogy, chemistry, physics, mathematics, zoology, and botany.

While still at high school he had evidently developed a keen interest in mineralogy. During his summer vacation in 1904 he discovered an occurrence of fine quartz crystals in fissures in schist at Fefor in Gudbrandsdalen in the Norwegian mountains. These crystals had strong pyroluminescence (emission of light when heated to 200–300°C), which is an unusual phenomenon, since most quartz crystals from other localities which he tested showed little or no pyroluminescence. He evidently discussed his research with Professors Brøgger and Hiortdahl,[5] who provided him with additional material from the university and their private collections. He made extensive experiments to explain pyroluminescence and its relationship to triboluminescence and phosphorescence. He exposed quartz crystals, which had their pyroluminescence destroyed by heating, to sunlight, arc light, electric sparks, and X-rays, none of which regenerated the pyroluminescence. However, the pyroluminescence was restored by exposing the crystals to the radiation from radium bromide (loaned by Professor Birkeland,[6] of Birkeland–Eyde fame). Professor Brøgger presented Goldschmidt's research on the pyroluminescence of quartz at the Academy of Sciences meeting on February 9, 1906, and his 19-page paper "Die Pyrolumineszenz des Quarzes" was published in the Academy's journal later that year.

During the spring of 1906 he travelled extensively on the continent with his parents. He observed the spectacular eruption of Vesuvius in April. Writing to Professor Brøgger on May 24, he described his visit to the villages of Ottajano and San Giuseppe, which had been destroyed by the fall of volcanic ash. He collected rocks and minerals from Vesuvius which he shipped to Brøgger from Naples. He apparently had promised Brøgger to collect rocks for him in Bohemia, but was unable to do this because his parents became ill and he had to remain with them in Baden–Baden. In the same letter he informed Brøgger that he had been invited on an excursion to the Eifel, and offered to collect samples from the classic volcanic localities. Thus already in 1906 there was a close, friendly relationship between Goldschmidt and Brøgger, a relationship that was life-long. Goldschmidt had enormous respect for Brøgger as a person, a teacher, and a scientist. This respect was reciprocated by Brøgger, who early recognized his potential and did much to encourage its development.

In the summer of 1906 Goldschmidt worked for some weeks at the University of Freiburg, studying radioactivity with Professor G. Meyer. A fellow student was George de Hevesy,[7] and this acquaintanceship was to ripen into a lasting friendship.

3 / The Petrographic Years: 1907–1921

IN THE SPRING OF 1907 Goldschmidt began his first major research work, which was to result in his doctor's thesis. It was a 483–page monograph, entitled *Die Kontaktmetamorphose im Kristianiagebiet* (The Contact Metamorphism in the Kristiania Region). Published in 1911, it is a classic in the literature of geology.

The Oslo (Kristiania before 1925) region is a graben, a trough of younger rocks downfaulted between blocks of ancient Precambrian[1] gneisses. The oldest rocks in the graben are Lower Paleozoic (Cambrian-Silurian) shales, marls (calcareous shales), and limestones. These rocks were later intruded in Permian times by magmas, mainly syenites of different types. The intrusions heated the intruded rocks and brought about the extensive mineralogical changes comprised under the term "contact metamorphism." The resulting rock types are known collectively as hornfels. These rocks are extremely fine-grained and their mineral identification demanded a remarkable facility in the use of the microscope, a facility which Goldschmidt demonstrated throughout his life.

As Goldschmidt pointed out in the introduction to his monograph, the Oslo region was uniquely suited to this investigation. The geology had been studied and recently remapped by his teacher, W. C. Brøgger, with many coworkers; the contacts between the intrusives and the sedimentary rocks were well exposed in many places, in quarries, mines, and road and railway cuttings; and extensive collections of minerals from the contact zones were available for study in the university collections. For the first couple of years he concentrated on field work and on the investigation of the mineralogy of the contact zones; more than half of his thesis consists of detailed descriptions of 83 minerals. As he noted in his preface:

> "The greater part of the geological work was completed in the summer of 1909. . . . Since the spring of 1910, after I had established the mineralogical composition of the hornfelses, I devoted myself to the problem, which I had considered important since the

beginning of my research, namely the principles governing the mineral associations in these rocks" (translated from German).

As a result of his comprehensive studies Goldschmidt concluded that the mineral association of a specific hornfels is completely determined by the composition of the original material and the pressure and temperature of recrystallization. Considering the series of rocks ranging in composition from shales to limestones, the resulting hornfelses will exhibit a corresponding range of definite and predictable mineral compositions. On this basis he established a series of ten hornfels classes, characterized by specific mineral associations as follows (in order of increasing calcium content):

1. andalusite-cordierite
2. andalusite-cordierite-plagioclase
3. cordierite-plagioclase
4. cordierite-plagioclase-hypersthene
5. hypersthene-plagioclase
6. hypersthene-plagioclase-diopside
7. plagioclase-diopside
8. plagioclase-diopside-grossular
9. diopside-grossular
10. diopside-grossular-wollastonite

Along with these specific minerals, the rocks also contained quartz, orthoclase, and often biotite.

This sequence can be visualized in terms of an *ACF* diagram (Fig. 1), in which *A* signifies Al_2O_3, *C* is CaO, and *F* is (Mg,Fe)O; the corners of the triangular diagram represent 100% of the specific component; the diagram presupposes the presence of excess SiO_2, present as quartz. *ACF* diagrams were devised by the Finnish geologist Pentti Eskola[2] while working with Goldschmidt in Oslo, 1919–20.

The *ACF* diagram not only illustrates the observed mineral associations, it also provides the explanation for the non–occurrence of conceivable mineral associations. Thus andalusite and hypersthene do not occur together, since they would react to give the intermediate compound cordierite; similarly wollastonite and anorthite are incompatible, reacting to form grossular.

In his doctoral thesis Goldschmidt used the data on the mineralogical compositions of this sequence of hornfels to arrive at a general principle, which he called the *Mineralogical Phase Rule*. In this he was probably influenced by his father, earlier a colleague of van't Hoff, who had applied the Phase Rule to the study of salt deposits. The Phase Rule is a chemical principle enunciated by the American chemist J. Willard Gibbs in 1877. It is usually stated in the form of an equation $P + F = C + 2$, or, in any system at equilibrium the number of phases, *P*, (mineral species in the case of rocks) plus the number of degrees of freedom, *F*, (temperature, pressure, chemical composition) is equal

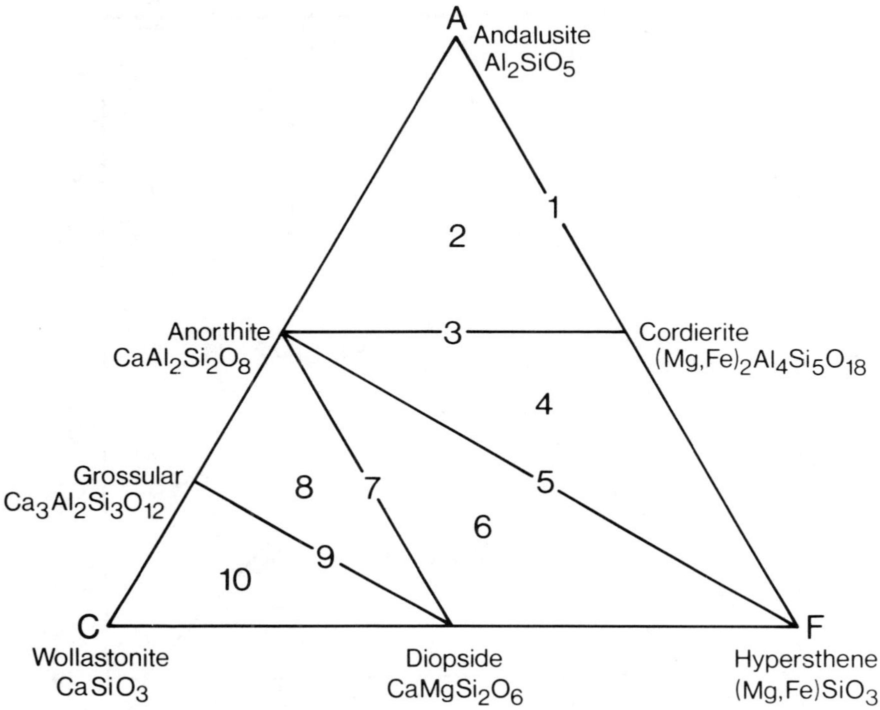

FIG. 1. ACF diagram illustrating the sequential mineral composition of Goldschmidt's ten classes of hornfels (Eskola, *Norsk Geol. Tidssk.*, vol. 6, p. 158, 1920). See text for additional details.

to the number of components, C, plus two. A simple example is the one–component system Al_2SiO_5, which occurs naturally as the minerals kyanite, andalusite, and sillimanite. Each of these minerals is stable over a range of temperatures and pressures (Fig. 2). The phase rule predicts that the three minerals can coexist in stable equilibrium only when $F = O$, i.e., both temperature and pressure are fixed. This is called the triple point, which for Al_2SiO_5 is at 500°C and 3.8 kilobars pressure, and it corresponds to a depth of about 13 km in the Earth's crust. But rocks crystallize over a range of temperatures and pressures, i.e., $F = 2$, and under these conditions $P = C$, i.e., only one of the three possible minerals will be present. This is borne out by experience, since rocks containing more than one of these minerals are very rare and usually show evidence of disequilibrium.

This is an example of Goldschmidt's *Mineralogical Phase Rule*, which states that in any rock the number of minerals will not exceed the number of components. This is a powerful tool in petrology in limiting the number of minerals present in a rock. Most rocks can be considered as within an eight–component system of O, Si, Al, Fe, Mg, Ca, Na, and K; these eight elements usually make up 99% of the common rocks. Seldom, if ever, will eight minerals occur in this sys-

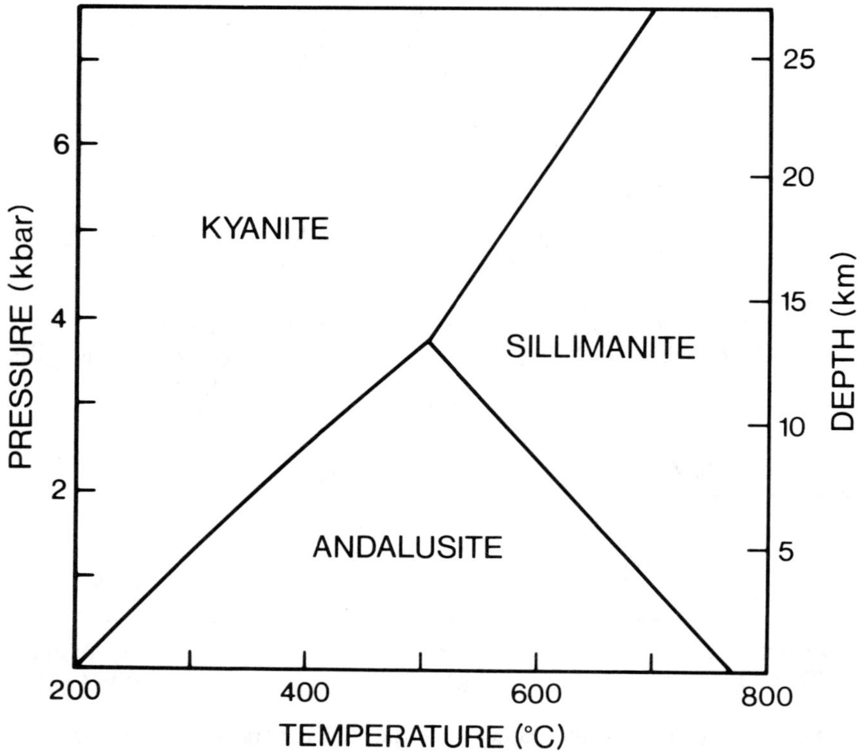

FIG. 2. Temperature–pressure relations in the Al_2SiO_5 system.

tem; several of these elements replace each other atom for atom in minerals, for example, Na and Ca in plagioclase feldspar, and Fe and Mg in ferromagnesian minerals, and are thus not truly independent components.

The Phase Rule has been of great utility in setting limits to the temperatures and pressures of rock formation. Goldschmidt initiated this in 1912 with his evaluation of this reaction:

$$\begin{array}{ccccccc} CaCO_3 & + & SiO_2 & = & CaSiO_3 & + & CO_2 \\ \text{calcite} & + & \text{quartz} & = & \text{wollastonite} & + & CO_2 \end{array}$$

a common metamorphic reaction whereby a siliceous limestone is converted into wollastonite. Using the limited data available at that time he determined a pressure–temperature curve for this reaction (Fig. 3). The curve for the equilibrium partial pressure of CO_2 divides the field into two parts, one of which (the lower part of the diagram) corresponds to the coexistence of calcite and quartz, and the other (the upper part of the diagram) to the field of wollastonite. Many analogous diagrams have since been determined and utilized in geothermometry and geobarometry (the determination of temperatures and pressures of reactions in the Earth's crust).

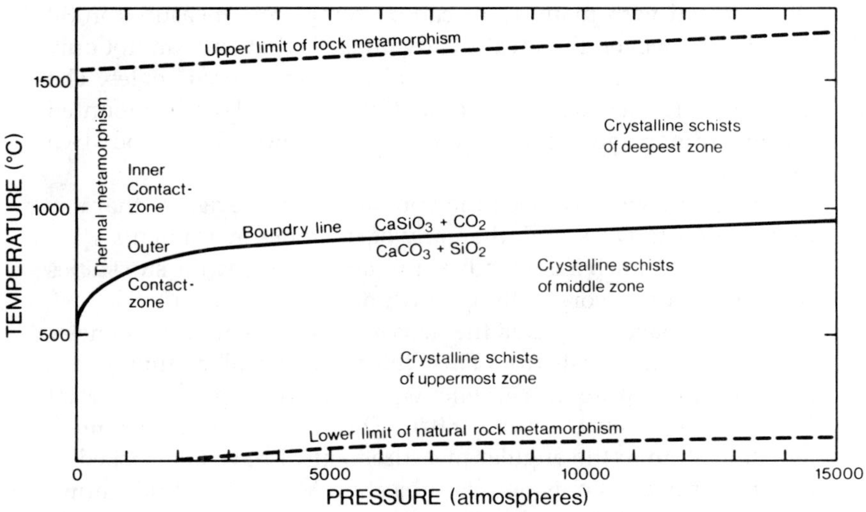

FIG. 3. Temperature–pressure relations in the $CaCO_3$–$CaSiO_3$–SiO_2 system (Goldschmidt, *Vidensk. Skrifter. I. Math.-naturv. klasse,* No. 22, p. 6, 1912).

While working towards his doctoral degree Goldschmidt took every opportunity to broaden his knowledge of geology and mineralogy. In the summer and autumn he did field mapping as an assistant on Norges Geologiske Undersøkelse (the Geological Survey of Norway). He spent the winter of 1908–09 at the University of Vienna, studying the techniques of optical mineralogy with Professor F. Becke.[3] In December, 1909, he received a university scholarship in Oslo in mineralogy and petrology; this involved some teaching responsibilities, and from then on he lectured in these subjects, and led field trips with the students. After obtaining his doctor's degree in May, 1911, he spent the winter of 1911–12 with Professor Paul Groth,[4] the foremost crystallographer of the time, at the University of Munich.

Professor Brøgger was concerned about his protégé's tendency to overwork, and evidently wrote to him about this when Goldschmidt was in Munich. Goldschmidt replied on February 17, 1912:

> "Many thanks for your friendly letter, which I received today. I was really ashamed when I read it, and I promise to improve. I must admit that I may have worked excessively in January, because it was necessary to complete the crystal measurements at low temperatures. It was rather trying work, and I am glad it is finished.
>
> Now, however, the manuscript is in the hands of Professor Groth, and I have been able to relax during the last few weeks. My only work is to try out the new four-circle goniometer which has been delivered by Fuess [a German instrument company]. I must say that it is a great pleasure to work with this instrument. Crystal measurements now take fewer hours than previously days, and a high degree of precision is attainable. For example, a mono-

clinic crystal with many faces can be completely measured in 30 minutes. However, the greatest advantage is that one can not only measure the interfacial angles, but also the interzonal angles. The instrument is not expensive; it costs 150 marks and can be mounted on an ordinary goniometer, and can be removed in about two minutes.

Everyone who has seen the goniometer in use agrees that it is much better to work with than any of the older instruments.

Here in Munich it is already springtime; each day the sun shines warmer, and we work with open windows.

When I have completed the descriptions of the new goniometer and the quartz crystals from Finse [Norway], I shall continue with the low temperature measurements. I am going to measure crystal hardness and cleavability at −180°C. This research should be much less difficult than the angular measurements. I have been asked to lecture to the physicists on this subject on March 6. All this, however, is in the future, and at present I work only a few hours each day. The best indication that I am not overstrained is that I am sleeping well. As soon as I work too much I sleep badly, and so far that has not happened this winter. You can be quite sure that I am not sacrificing my health on account of my work.

Working here at the Institute is enjoyable in all respects, and Groth's assistants are all nice people, with whom it is a pleasure to collaborate. Professor Groth is always very friendly towards me; he sends you his warmest regards.

The collection of contact metamorphic rocks from the Kristiania region is already on exhibition; I am to select a corresponding collection of Alpine rocks, which will be forwarded during the spring.

I hope you are well. Thank you again for your friendly concern" (translated from Norwegian).

On his return from Munich Goldschmidt was appointed docent (assistant professor) in mineralogy and petrology at the University of Oslo from July 1, 1912.

After his experience with contact metamorphism in the Oslo region, he turned his attention to the wider and more difficult field of regional metamorphism, as it is developed in the cores of mountain ranges, such as those of southern Norway. Beginning in the summer of 1911 and continuing until late in 1916, he mapped large areas of the mountain country between Stavanger and Trondheim. In this work he was assisted by Mimi Johnson[5] and Endre Berner[6] (Plate 5).

In March 1914 Goldschmidt was elected to the Videnskaps–Akademi (the Norwegian equivalent of the Royal Society of London), one of the youngest persons so honored. Some of the Academy opposed this, on the grounds that there were older scientists equally qualified. Brøgger carried the day for Goldschmidt by saying:[7] "I do not hesitate to say that I consider Victor Goldschmidt the most outstanding Norwegian mineralogist of today—myself not excepted!"

In 1914 the professorship in mineralogy at the University of Stockholm became vacant, and Goldschmidt applied for it. Of the four applicants he was placed first and was offered the position—a remarkable achievement in view of his youthful 26 years! The University of Oslo, not wishing to lose this rising star, prevailed upon the government to establish a Mineralogical Institute with a personal professorship for him. Professor Brøgger, who was active politically at that time, was probably largely responsible for this outcome. On September 28, 1914, Goldschmidt gave his inaugural lecture entitled, "On the Problems of Mineralogy." The following excerpt outlines his interests at that time.[8]

> "What are the conditions for thermodynamic equilibrium (in geological systems), and why is it that we find some minerals in one occurrence and not in another? The answer is that equilibrium conditions depend upon three independently variable factors. The first of these is represented by the bulk chemical composition of the existing minerals; it tells us which starting material the mineral-forming processes had to work with. The two other factors are the temperature and the pressure under which crystallization took place. These factors depend upon the geological conditions of the mineral occurrence. From one given starting material different minerals may form, depending on the temperature and pressure. In some cases we may from purely thermodynamic reasoning predict the temperature and pressure which must exist in order for certain mineral associations to be formed. One example of this, to which I once called attention, is the formation of calcium-rich silicates at the expense of calcium carbonate. This process can only take place within a certain pressure and temperature range. Thus, we have a method to determine the physical conditions, temperature and pressure, which have existed in the crust of the earth during mineral formation.
>
> It is of course of great interest to be able to determine the physical conditions under which an individual mineral has been formed. It is of much greater importance not to limit such investigations to one single occurrence, but to study thoroughly a sizeable area in order to investigate the temperature–pressure distribution during a certain geological era.
>
> Such an investigation, no doubt the first of its kind, is being performed by myself in the Norwegian mountain areas, from Ryfylke [near Stavanger] to Trondhjemsfjorden, in order to determine the temperature and pressure conditions in this part of the earth's crust during the formation of the Norwegian Caledonides at the beginning of the Devonian. The procedure of such a systematic investigation is as follows: rock specimens are collected a few kilometers apart and each specimen is studied by optical methods to recognize certain signs indicating the temperature and pressure

> which existed in the locality in question during the formation of the rock. The sum of all observations gives us a picture of the temperature–pressure distribution during the formation of a mountain chain."

The results of these investigations in the Norwegian mountains were published between 1912 and 1921 in a series of five monographs with the common title *Geologisch-petrographische Studien im Hochgebirge des südlichen Norwegen.* The first two are purely geological, on a Cambrian conglomerate at Finse, and on the deformation of the Precambrian basement during the Caledonian orogeny. The third describes the metamorphic rocks of the Trondheim region, accompanied by a map showing the distribution of the various grades of metamorphism, as determined by the occurrence of distinctive index minerals such as chlorite, biotite, and garnet.[9] He compared and contrasted the relationship between the contact metamorphism of the Oslo region and the regional metamorphism in the Trondheim area. The fourth monograph deals with the igneous rocks associated with the metamorphic rocks in the Norwegian mountains. He introduced the concept of *Gesteins–Stamm* (*rock kindred*) to denote a series of genetically–related igneous rocks within a geologically limited region. He distinguished three such kindreds in the Norwegian mountains, each with characteristic features. These were the "green rock kindred" (basic lavas and intrusives of the Trondheim region and on the west coast), the "Bergen–Jotun kindred" (the anorthosites and related rocks in Jotunheim), and the "opdalite–trondhjemite kindred" (opdalite is a name introduced by Goldschmidt for a rock whose composition corresponds closely to the mean composition of igneous rocks, and trondhjemite is his name for a white plagioclase–rich granite, now recognized world–wide in many similar regions). The final monograph, published several years after the completion of the field work, describes the injection metamorphism of the Stavanger region. Here are found mixed rocks of metamorphic and igneous material in a form suggesting that granitic magma was injected between layers of metamorphic rock, and he called these rocks injection gneisses (now generally known as migmatites, a name introduced for similar rocks in southern Finland). The nature and origin of migmatites is still a controversial field of great geological interest.[10]

Goldschmidt expounded on his concepts of the evolution of igneous rocks in a short monograph entitled *Stammestypen der Eruptivgeseine,* published in 1922. This is illustrated in Fig. 4. Discussing the progressive crystallization of a magma he wrote:

> "Among the first products of separation are minerals with low solubility in the silicate magma and high melting points, such as chromite and olivine, then follow other iron ores, together with the bulk of pyroxenes and plagioclase feldspars; amphibole and biotite succeed the pyroxenes, the calcic plagioclases are followed by sodic

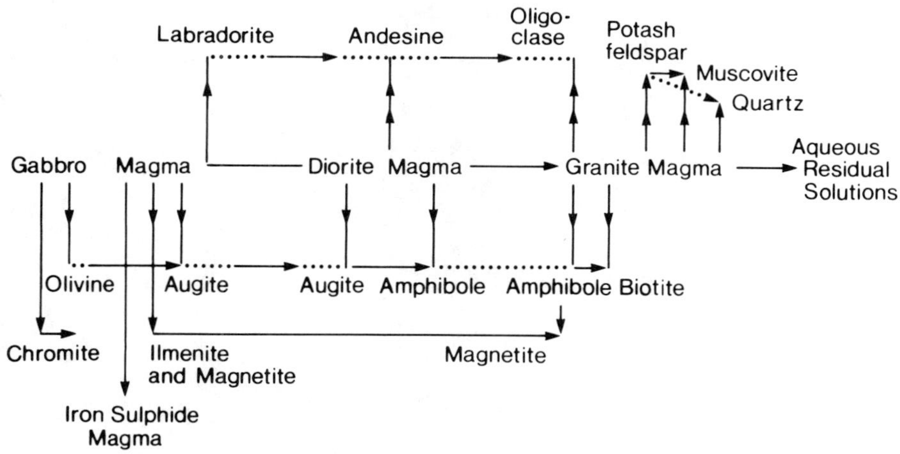

FIG. 4. The evolution of igneous magmas through fractional crystallization (Goldschmidt, *Vidensk. Skrifter. I. Math.–naturv. klasse,* No. 10, p. 6, 1922). The first-formed minerals (e.g., chromite) and rocks (e.g., gabbro) are on the left, and the final crystallization products (aqueous residual solutions) on the right.

> ones and by potash feldspar, with quartz and muscovite. Finally, only pegmatitic magmas and aqueous solutions remain as a kind of mother liquor of the magma, from which crystallize coarse pegmatitic dikes and hydrothermal mineral veins, and from which also gaseous constituents may escape, giving rise to pneumatolitic minerals" (translated from German).

In a remarkable example of parallel thinking, the same concepts were published simultaneously by N. L. Bowen in his paper entitled "The reaction principle in petrogenesis" (*J. Geol.*, vol. 30, p. 177–196, 1922). Bowen's thesis was based largely on experimental work on the crystallization of silicate melts, which were carried out by his colleagues and himself at the Geophysical Laboratory in Washington, D.C., during the previous decade. In his monograph Goldschmidt refers appreciatively to Bowen's work.

In June 1922 Goldschmidt took an international group of distinguished geologists on a tour of the Stavanger area to demonstrate his injection gneisses and other features. The group included the following: F. Becke (Austria); C. E. Tilley (England); L. H. Borgström and H. G. Backlund (Finland); O. Andersen, O. A. Broch, C. Bugge, S. Foslie, C. F. and N. H. Kolderup, I. Oftedal, and Th. Vogt (Norway); and H. von Eckermann, A. Gavelin, P. J. Holmquist, H. Johansson, and P. D. Quensel (Sweden). An album of photographs from this excursion is preserved in the Goldschmidt Archives and a second copy in the Geological Museum in Oslo. In his instructions to the participants he recommended that they bring with them a half–bottle of brandy for medicinal use during the excursion, since at that time Norway was experimenting with a modified form of Prohibition.

4 / *The Raw Materials Laboratory*

WORLD WAR I WAS to produce a profound change in Goldschmidt's scientific development. As the war progressed, Norway was largely cut off from overseas, and sources of raw materials, previously imported, were badly needed. In 1916 he was investigating the possibility of replacing imported bauxite with Norwegian clay or feldspar as raw material for aluminum production. In November 1917 the Norwegian government decided to finance research into the mineral resources of the country, and established a Commission for Raw Materials, with Goldschmidt as Chairman. On his initiative the commission approved the founding of Statens Råstofflaboratorium (the State Raw Materials Laboratory), which was set up in the newly built Geological Museum in the Oslo suburb of Tøyen. This gave the opportunity to obtain staffing and equipment for the new laboratory, financed by the Department of Industry and Commerce, thereby providing Goldschmidt with an institute independent of the university.[1]

One of the first fruits of this new organization was of much greater scientific than economic importance. In May 1918 Goldschmidt visited the Fen region in southern Norway to investigate the possible exploitation of the limestone known to occur there. He noted the presence in this limestone of a rare niobium mineral, pyrochlore, $(Na, Ca)_2Nb_2O_6(OH, F)$, and drew the conclusion that this was due to the local intrusion of sodium-rich igneous rocks. In August he spent a week in the area together with Professor Brøgger, during which they discovered and mapped a remarkable variety of unusual rocks. They completed the mapping during a further visit in May 1919, and Brøgger undertook the mineralogical and petrographical investigations. This resulted in Brøgger's classic 408-page monograph[2] *Das Fengebiet in Telemark, Norwegen*, a principal result being that the "limestone" was not a sedimentary rock, as had been thought, but had been intruded as a carbonate–rich melt. Carbonate–rich melts had been invoked by a few earlier petrologists, but never with the wealth of descriptive detail given by Brøgger. The Fen region is generally recognized as the type locality for carbonatites (the term introduced by Brøgger), as similar rocks in many parts of the world are now known.

Some of the Fen carbonatites contain considerable amounts of the phosphate mineral apatite. During World War II (until he left Norway in 1942) Goldschmidt had a research project, financed by the Agricultural Department, for the manufacture of phosphate fertilizer from this material. The importance of this research for Norwegian agriculture saved him, at least temporarily, from deportation to Poland in 1942. After World War II the carbonatites were mined for pyrochlore as a source of niobium, a metal needed in applications of atomic energy, and as a component of special steels.

The early work of the Raw Materials Laboratory concentrated on finding Norwegian sources of previously imported materials, and developing substitutes where possible. These included the possible production of aluminum from clays or anorthosite (a rock present in large deposits in Norway and containing 30% or more Al_2O_3); the use of potassium minerals, especially biotite, as a substitute for imported potassium fertilizers; the search for phosphate deposits; and the utilization of Norwegian ilmenite ($FeTiO_3$) in the preparation of titanium dioxide for use in paints. This led Goldschmidt to investigate more closely the chemical nature of economic minerals, and thus to the study of crystal chemistry. Fortuitously, this coincided with the development of new and powerful techniques for elucidating crystal structure. X–ray diffraction by solids had been discovered by Max von Laue[3] in 1912 and applied to the determination of crystal structures, especially by W. H.[4] and W. L. Bragg[5] in England. However, their procedures required the use of large well–formed crystals, sample preparation and data gathering were tedious and time–consuming, and World War I effectively suspended most research in this field. P. Debye and P. Scherrer in 1916 and A. W. Hull in 1917 independently developed the powder diffraction technique, which eliminated the need for good crystals and made data gathering relatively simple and rapid. Goldschmidt was to exploit this new technique brilliantly, and he used it to develop a new science of crystal chemistry.

J. D. Bernal,[6] a British physicist who specialized in crystal chemistry, recounts the following story of how Goldschmidt adopted X–ray diffraction as a research tool:[7]

> "The actual story as he told it to me—I don't know whether it is true or not, but it is a good Goldschmidt story—was that it all came about through a lawsuit. At the end of the first World War the then government of this country, feeling that it had to protect the British chemical industry which had been left so undeveloped before the war, produced a tariff on the import of organic chemicals. Among those referred to was calcium carbide. The chemical interests in this country claimed it was an organic chemical; the Norwegian government maintained it was not. A test case was to be fought in

London. Goldschmidt was briefed by the Norwegian government to be the expert witness on the subject. He certainly was not an organic chemist, but he read a textbook on the boat going across and set forth bravely for the law court. When he learned on arrival that the British government expert was Sir William Pope, he decided after all it would not do and he would have to fight him on another ground. Luckily he knew Sir William Bragg, so he took a crystal of calcium carbide and asked him to work out the structure (incidentally he got it completely wrong, but no one could know that at the time), and Goldschmidt fought the case on the basis that this was the structure of an inorganic crystal. There could be no argument against this on chemical grounds, and calcium carbide was finally exempted. The Norwegian government was naturally pleased and asked how it was done. He put it all down to X-rays and on the strength of this obtained an X-ray tube and a powder camera."

At this time Goldschmidt began recruiting a remarkable group of young research associates, all of whom were to have distinguished careers. In 1918 Ivar Oftedal[8] joined the museum staff as a student assistant and began an association with Goldschmidt which lasted throughout the latter's life; on his death Oftedal succeeded him as Professor of Mineralogy. Lars Thomassen[9] graduated as a chemical engineer in 1919, and from then until 1929 (except for two years at the California Institute of Technology, 1925–27) was Goldschmidt's principal research associate. He was responsible for the construction and operation of the X-ray equipment. In June 1921 Kristoffer Stenvik[10] joined the Raw Materials Laboratory as a technical assistant. He studied mineralogy, petrology, and chemistry at the university, and became Goldschmidt's right-hand man, especially in the development and utilization of olivine as a refractory. They complemented each other in a remarkable manner—Goldschmidt with his theoretical knowledge and fertile ideas, and Stenvik with his technical expertise.

The utilization of olivine is one of the great success stories of the Raw Materials Laboratory. Norway has large deposits of olivine, which are easily exploitable and readily accessible for shipping. Originally, Goldschmidt visualized olivine as a source of magnesium and magnesium compounds, since Norwegian olivine contains up to 50% MgO. However, for this purpose the extraction techniques were uneconomic, compared with the availability of magnesium from salt deposits and sea water. He then turned to the utilization of olivine as a refractory material, inspired, so he said, by observing the resistance of olivine xenoliths in basalt to melting and chemical attack, despite their immersion in molten lava. In 1925 he found that magnesium-rich olivine has a much higher melting point than previously thought, and began investigating its technical use as a refractory. During the 1930's he continued the early investigations, largely at his own expense, and obtained numerous patents in Europe, North America, and other countries, from which he eventually received considerable income. The great expansion in the commercial use of olivine came, however, after his death in 1947.

The year 1922 appears to have been a critical one in the development of Goldschmidt's concept of geochemistry and his plans for research thereon. He wrote two short but seminal papers, *Der Stoffwechsel der Erde* (The differentiation of the Earth), 25 pages, and *Geochemische Verteilungsgesetze der Elemente* (The laws of distribution of the chemical elements), 17 pages, the first of a series of nine monographs published from 1923 to 1938. In "Stoffwechsel" he outlined his ideas as follows:

> "It is conceivable that the original state of the Earth was a homogeneous or nearly homogeneous mixture of the chemical elements and their compounds. Today, however, the Earth is far removed from a homogeneous state. The material distribution within the Earth has by no means reached a final equilibrium state; we observe instead an active redistribution of matter and energy. The processes which have resulted in the inhomogenity of our planet and still contribute to the migration of material I would summarize in the expression "Der Stoffwechsel der Erde" " (p. 3, translated from German).

By analogy with the composition of meteorites (at that time generally believed to be fragments from a disrupted planet), which contain a nickel–iron alloy, iron sulfide (FeS), and silicate minerals, Goldschmidt postulated that the initial condition of the Earth was a molten ball surrounded by a gaseous atmosphere. The molten ball would have three immiscible liquids—nickel–iron, iron sulfide, and silicate. As a result of gravitational forces, the densest liquid, nickel–iron, would sink to form a core; the core would be surrounded by the sulfide liquid, possibly including some metallic oxides; and the silicate liquid, being the least dense, would form the mantle. This he termed the primary geochemical differentiation; the chemical elements would distribute themselves between the three liquid phases according to their affinity for metal, sulfide, or silicate; gases such as nitrogen and oxides of carbon would form the primitive atmosphere. The next step, the secondary geochemical differentiation as he called it, was the cooling and crystallization of the silicate liquid. The denser minerals, mainly iron–magnesium silicates, would tend to sink; the less dense minerals, mainly quartz and feldspars, would tend to float and eventually aggregate to form the continental crust. The crystalline minerals acted as sorting mechanisms for individual elements by accepting those that could be accommodated within the rigid structures and rejecting those either too large or too small.

On the basis of these considerations, Goldschmidt devised the geochemical classification of the elements into four groups: *siderophile* (those with an affinity to nickel–iron), *chalcophile* (those with sulfide affinity), *lithophile* (those with silicate affinity), and *atmophile* (those normally present as gases). The geochemical character of an element is clearly related to its position in the Periodic Table,[11] as is seen in Table 1.

Table 1. Geochemical classification of the elements in relation to the Periodic System (after Goldschmidt, *Vidensk. Skrifter. I. Math.–naturv. klasse*, No. 3, p. 5, 1923).

Atmophile: N
Lithophile: Na
Chalcophile: Zn
Siderophile: Fe

H																	He
Li	Be											B	C	N	O	F	Ne
Na	Mg											Al	Si	P	S	Cl	Ar
K	Ca	Sc	Ti	V	Cr	Mn	Fe	Co	Ni	Cu	Zn	Ga	Ge	As	Se	Br	Kr
Rb	Sr	Y	Zr	Nb	Mo		Ru	Rh	Pd	Ag	Cd	In	Sn	Sb	Te	I	Xe
Cs	Ba	La-Lu	Hf	Ta	W	Re	Os	Ir	Pt	Au	Hg	Tl	Pb	Bi			
			Th		U												

When Goldschmidt proposed his geochemical classification of the elements in 1923, few quantitative data were available on which to base his ideas. The geochemical nature of an element could, of course, be established by measuring its distribution between three liquid phases of iron, iron sulfide, and silicate. He recognized the difficulty in carrying out these measurements in the laboratory, but remarked that meteorites provide us with such an experiment in a fossilized condition. Many meteorites consist of nickel–iron, troilite (FeS), and silicate, all of which probably solidified from a melt. The distribution of a specific element between these three phases would be established when the system was liquid and can be determined by mechanically separating these phases and analyzing them individually. From such analyses the partition of the elements between metal, silicate, and sulfide can easily be calculated. In the following years Goldschmidt and his co-workers made many measurements of the content of various elements in nickel–iron, troilite, and silicate of meteorites. He realized that these measurements also provided the data to establish the cosmic or absolute abundances of the elements, since meteorites provided an unaltered sample of the material which coalesced to form our Solar System. He used these data to compile the first table of the absolute abundances of the elements, which he published in 1938, and which provided the basis for theories of the origin of the elements. At a time when the study of meteorites was a scientific backwater largely confined to description and classification, Goldschmidt's perception of the fundamental significance of meteorite compositions for geochemistry is another example of his scientific genius.

In 1922 Goldschmidt established the conceptual basis for his new geochemistry. It now remained to flesh out this skeleton with experimental data on the distribution of the chemical element in minerals and rocks. This required two sorts of information: a knowledge of the crystal structures of the minerals, which controlled the distribution of the elements, and the analysis of these minerals for their elemental composition, both major and minor. The time was ripe, because of the availability of newly developed analytical techniques. X–ray powder diffraction provided a rapid method for the determination and comparison of crystal structures. For the elemental analysis of minerals a new technique had just become available. In 1921 Assar Hadding,[12] a Swedish scientist and friend of Goldschmidt, had designed an X–ray spectrograph specifically for the analysis of minerals. X–ray spectra show a relatively simple pattern of lines in a regular progression from one element to the next. An enormous advantage over classical chemical analysis was that it was unnecessary to separate out the individual elements—the mineral grains were mounted directly on the X–ray target, and a single spectrum showed the lines for each of the elements present, and the intensities of the lines were approximately proportional to the amount present.

Goldschmidt's appreciation of the value of Hadding's X–ray spectrograph for his geochemical researches is expressed in a letter dated January 19, 1924:

> "Dear friend. . . . Every day I am more and more convinced of the excellence of the X–ray spectral analysis technique for our work on the laws of distribution of the chemical elements. No other method is capable of producing so rapidly and exactly the analyses needed for my work on this problem. My institute has never acquired a more useful piece of equipment. . . . You can be justifiably proud to have developed this method, which opens a whole new field in mineralogical research" (translated from Norwegian).

In March 1922 Thomassen built an X–ray spectrograph to Hadding's design. It was destined to be used for many investigations in Goldschmidt's institute during the next few years, but its first application was the search for element 72.

5 / Intermezzo: The Search for Element 72[1]

ALTHOUGH THE ENUNCIATION of the Periodic Law of the Elements by D. I. Mendeleev (1834–1907) in 1869 brought system to the arrangement of the chemical elements and enabled the prediction of some yet to be discovered, it could not define the total number, largely because of the unknown number of rare earth elements falling between barium (atomic weight 137.33) and tantalum (atomic weight 180.95). Since barium is bivalent, the rare earths trivalent, and tantalum pentavalent, hindsight predicts that a quadrivalent element should exist between the rare earths and tantalum, but this does not seem to have been recognized.

The problem of the total number of elements was solved by H. G. J. Moseley (1888–1915). After graduation from Oxford in 1910 he joined Rutherford's research group at the University of Manchester. Work there was concentrated on problems of radioactivity and atomic structure, but the discovery of the diffraction of X-rays by crystals by Max von Laue in 1912 turned Moseley's attention to this new field. He used the diffraction technique to determine the wave lengths of X-rays produced by bombarding targets of specific elements with electrons. The results were published in two papers in the *Philosophical Magazine,* in December 1913 and May 1914. The second paper contained a listing of the elements from aluminum with atomic number (AN) 13 through gold (AN 79), and he stated "Every element from aluminum to gold is characterized by an integer N which determines its X-ray spectrum; this integer is the atomic number of the element." He identified three gaps in the sequence of elements, at atomic numbers 43, 61, and 75.

At that time it was believed that element 72 had been discovered in 1911 by Georges Urbain (1872–1938), a French chemist who spent many years separating and characterizing rare earth elements; he named it celtium. Urbain was impressed by Moseley's discovery and visited him in June 1914, bringing with him the precious separates supposedly containing celtium. In two days Moseley produced the

X–ray spectrum. Alas, the spectrum showed only the lines characteristic of ytterbium (AN 70) and lutetium (AN 71). Element 72 was still missing.

This was Moseley's last scientific work. Shortly thereafter he sailed for Australia to attend a meeting of the British Association for the Advancement of Science in Melbourne. World War I broke out during the meeting; Moseley hurried back to England to enlist and was posted to the Royal Engineers. Rutherford deplored the acceptance of such scientific talent into the armed forces and tried to have Moseley transferred to the National Physical Laboratory. The transfer was approved but not soon enough. Moseley had been sent to Gallipoli, where a Turkish bullet ended his life on August 10, 1915.

World War I and its aftermath largely suspended scientific research not directly related to the war effort. The search for element 72 was taken up again in 1922, this time in Copenhagen and in Oslo. The Danish physicist Niels Bohr (1885–1962), on the basis of his theory of atomic structure, predicted that this element could not be a rare earth, but must be quadrivalent and an analog of titanium and zirconium. During the summer of 1922 George de Hevesy, working in his institute, became interested in geochemistry and decided to look for this elusive element in zirconium minerals.

Hevesy had been a researcher in Rutherford's institute at the same time as Moseley, and they had become good friends. Indeed, Moseley invited Hevesy to join him in the summer of 1914 and work together on X–ray spectroscopy, an invitation postponed by his trip to Australia and terminated by the outbreak of war. Now Hevesy teamed up with Dirk Coster, a Dutch physicist who came to Bohr's institute in September 1922 to build an X–ray spectrograph. Hevesy obtained specimens of zirconium minerals from the Mineralogical Museum in Copenhagen. He treated the mineral zircon ($ZrSiO_4$) with boiling acid to remove the soluble constituents; the insoluble residue containing zirconium, and perhaps element 72, was fixed on the copper anticathode of the spectrograph. The first exposure taken by Coster showed spectral lines in the positions predicted for the missing element! Hevesy commented:[2] "The appliances at our disposal at that time were very primitive and it took us several weeks to secure conclusive evidence of the discovery of the element." In a letter to *Nature* dated January 2, 1923 and published in the January 20 issue Coster and Hevesy announced the discovery of element 72 in zircon and named it hafnium, from the Latin name for Copenhagen.

In 1922 Goldschmidt and Thomassen were using their newly built X–ray spectrograph for geochemical research, and were also looking for element 72. In the first issue of *Norsk Geologisk Tidsskrift* for 1923 they published an article dated January 31, 1923 entitled "Das Vorkommen [occurrence] des Elementes No. 72 [Hafnium] im Malacon und Alvit" (malacon and alvite are varieties of zircon).

Being beaten to the publication of the discovery of element 72 by only 29 days was a great disappointment to Goldschmidt. For years thereafter he devoted much time and effort in the search for the then unknown element 61. Since it is one of the rare earth elements, he analysed any new rare earth mineral received by the museum. By the 1930's, however, he decided that this element probably had no stable isotope and did not occur naturally. This was confirmed in the 1940's, when the element was identified as one of the products of the fission of uranium, and named promethium (Prometheus, according to mythology, stole fire from heaven); several isotopes are known, the longest–lived having a half–life of 17.7 years.

Goldschmidt sent a sample of his alvite to Hevesy, who confirmed its high hafnium content (16% HfO_2, whereas in most zircons HfO_2 ranged from 1.3 to 6%). Hevesy included this result among others in an invited lecture at the 1923 annual meeting of the German Chemical Society in Berlin. When Goldschmidt learned from Norwegian colleagues present at the meeting that Hevesy had omitted reference to his gift of this particular specimen, he, as Hevesy says, "cut off all connection." Friendly relations, however, were restored when proper reference appeared in the published proceedings. Sensitivity on professional issues, however, remained; a new incident occurred in 1925. This is suggested by the following letter from Goldschmidt, presumably provoked by a complaint from Hevesy about institutional priority distortions apparent from a Danish newspaper account of a lecture given by Goldschmidt:[3]

> "Dear Colleague, Through Prof. Størmer I just received the Danish "Nationaltidende" [Copenhagen newspaper] of February 3. On page 6 is a note about the lecture which I have given, and which can give the erroneous impression that I have ascribed to myself results which are due to others, especially also to your Institute. I have self–evidently in my lecture drawn full attention to the fact that the quantitative X–ray spectral analysis has been used first in Copenhagen, and this is also expressly indicated in the sole Norwegian newspaper report. As you will see from this report, which I permit myself to attach, I have in no way tried to ascribe to myself results that are due to other institutions. That "Aftenposten" [Oslo newspaper] has brought some unfortunate superlatives also in connection with my person is of course to be regretted, but I will under any circumstances place great value on your seeing the article so that the impression shall not be made in Copenhagen that I will steal well deserved fame from Bohr's Institute. If you would regard it as necessary or desirable, I wish to ask you to present the Aftenposten article in original to the Nationaltidende. I assume that the Nationaltidende note must concern a misinterpreted review of the article by cable or telephone. The lecture which I gave was in reality concerned especially with quantitative spectral

> analysis and determination of the variability of sensitivity in different spectral regions, and a review of Thomassen's experiences in qualitative X–ray spectrography. Further I also showed the first spectrograms which were taken with diamond crystals. If you are interested, alvite can perhaps now be obtained with 7–8% hafnium oxide from a new deposit which has been found this winter. With best regards, Yours sincerely, V. M. Goldschmidt."

Despite its late discovery, hafnium is not an exceedingly rare element.[4] In the Earth's crust it is more abundant than the familiar elements arsenic and tin. Goldschmidt referred to it as a *camouflaged element,* which because of its great similarity to a more abundant element forms no minerals of its own, but is always dispersed in small amounts in the minerals of the more abundant element.

In March 1940 I had the privilege of meeting Hevesy over dinner in Goldschmidt's home, where he was staying as a house guest. He spoke fascinatingly of the work he was doing on the use of radioactive tracers in biological research, and when I expressed interest he invited me to be a guest at his institute in Copenhagen during the summer. In April, however, the Germans invaded Denmark and Norway, effectively negating that attractive invitation!

6 / *Crystal Chemistry and Geochemistry in Oslo: 1922–1929*

GOLDSCHMIDT DESCRIBED HIS plans for the Raw Materials Laboratory in the post–World War I period as follows:[1]

> "After the end of the first world war, when a new economic situation had developed, I recommended the continuance of research work along more general lines on raw minerals materials and the study of general scientific problems closely connected with the utilization of mineral raw materials. Among the problems proposed here there was one of outstanding importance, i.e. to find the general laws and principles which underlie the frequency and distribution of the various chemical elements in nature—the basic problem of geochemistry. I proposed to attack the problem from the view point of atomic physics and atomic chemistry, and to find out the relationships between the geochemical distribution of the various elements and the measurable properties of their atoms and ions"(original English).

Concurrently with their search for element 72, Goldschmidt and Thomassen began an extensive research on the abundance and distribution of the rare earth elements: yttrium (AN 39) and the sequence from lanthanum (AN 57) to lutetium (AN 71); Goldschmidt introduced the term *lanthanides* for the latter sequence. Information in this field was almost totally lacking, because of the extreme difficulty in separating the individual elements by the classical methods of chemical analysis. Their X–ray spectra, however, showed the relatively simple pattern of lines in a regular progression from one element to the next. An enormous advantage was that it was unnecessary to separate out the individual elements—a single spectrum showed the lines for each of the elements present, and the intensities of the lines were approximately proportional to the amount present. Goldschmidt and his research assistants separated the rare earths from a large number of different minerals, and Thomassen analyzed these separates with his

X–ray spectrograph. On the basis of these results they were able to estimate the relative abundances of these elements, as shown in Fig. 5, where they are normalized to the homologous element yttrium having an arbitrary abundance of 100. The figure confirms that cerium is the most abundant lanthanide, followed by neodymium, the remaining elements being relatively rare. The most remarkable feature is the sawtooth pattern of the abundances. This had been foreshadowed by Oddo in 1914 and independently by Harkins in 1917, in what is now known as the Oddo–Harkins rule. They stated it in the form that odd–numbered elements tended to be less abundant than even–numbered elements. On the basis of their results, Goldschmidt and Thomassen refined the rule, as follows: "Elements of odd atomic number are less abundant than their immediate neighbors of even atomic number."

They found no evidence for the presence of element 61 in any mineral. Nevertheless, Goldschmidt continued to search for it, as evidenced by his correspondence with Assar Hadding. Hadding to Goldschmidt (March 10, 1926): "How goes your search for 61?"

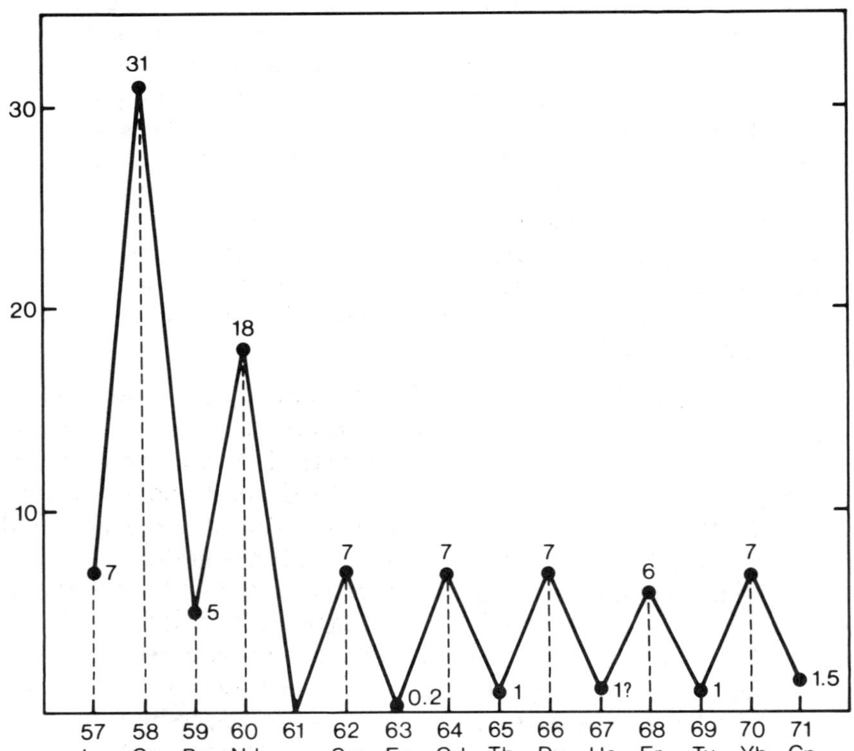

FIG. 5. The relative abundances of the lanthanides, referred to yttrium = 100 (Goldschimdt and Thomassen, *Geochemische Verteilungsgesetze der Elemente III*, p. 49, 1924). Note: Element 71 Cp (cassiopeium) is now Lu (lutetium).

Goldschmidt to Hadding (March 11, 1926): "I am still trying to find 61." He eventually decided that 61 did not exist in nature as a stable element. Goldschmidt and Thomassen also commented on the exceptionally low abundance of europium, even compared to the other odd–numbered elements. In 1926 Goldschmidt showed that this was due to the fact that under reducing conditions europium behaved differently from other rare earth elements in that it was present in the bivalent state; the bivalent europium ion has a considerably larger radius (1.17Å) than the trivalent ion (0.96Å) and is therefore excluded from the common rare earth minerals but accepted in calcium minerals. This phenomenon is now widely recognized and termed the *europium anomaly;* it is much used in the interpretation of the evolution of igneous rocks, especially those from the Moon, which crystallized under strongly reducing conditions.

During 1924 Goldschmidt's research tempo was greatly increased by the addition of three young scientific assistants paid from Raw Materials Laboratory funds: Tom. Barth, Gulbrand Lunde, and William Zachariasen,[2] all of whom had notable careers. The first fruits were the determination of the crystal structures of the oxides of the rare earth elements, published in 1925 as *Geochemische Verteilungsgesetze der Elemente IV*, with co–authors Ulrich[3] (research fellow from Prague) and Barth. In the introduction Goldschmidt acknowledges the work of Lunde in the preparation of the oxides, and Zachariasen in the interpretation of the X–ray data. Goldschmidt had a unique collection of purified rare earth compounds, the gift of his friends George de Hevesy and Auer von Welsbach.[4]

One of the remarkable results of this research was the discovery that the unit cell dimensions of the lanthanide oxides decrease regularly with increasing atomic number of the element (Fig. 6). This is contrary to the normal increase in unit cell dimensions with increase in atomic number—for example the cell dimensions of Sc_2O_3 (AN 21) and Y_2O_3 (AN 39) are 9.79Å and 10.60Å, respectively. These results demonstrated that the sizes of the lanthanide ions must decrease with increasing atomic number, the reason being that the orbital electrons needed to balance the increasing number of protons in the atomic nucleus are added not to the outer shell, but to an inner shell, and the effective radius of the ion is thereby diminished. As a result, the radius of yttrium falls near those of dysprosium and holmium, which explains the previously observed division of the lanthanides into the cerium group (those occurring naturally in cerium minerals, essentially La—Sm) and the yttrium group (those occurring in yttrium minerals, essentially Eu—Lu). It also explains why hafnium (AN 72) has essentially the same radius as zirconium (AN 40), and tantalum (AN 73) as niobium (AN 41).

Goldschmidt discussed this phenomenon at length in *Geochemische Verteilungsgesetze der Elemente V* (1925) and termed it the *lanthanide contraction.*

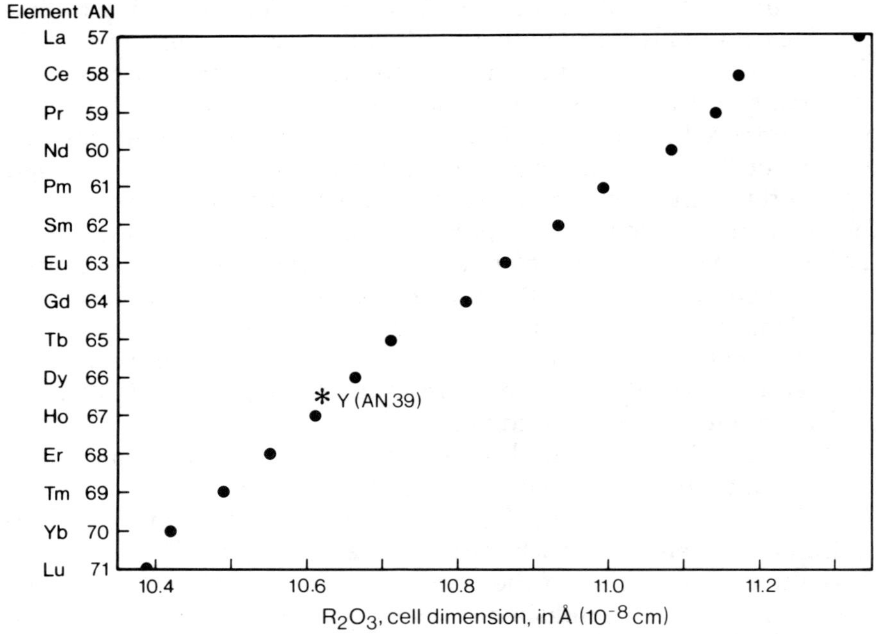

FIG. 6. Unit cell dimensions of cubic lanthanide oxides, illustrating the lanthanide contraction (from Goldschmidt, Ulrich, and Barth, *Geochemische Verteilungsgesetze der Elemente IV,* 1925, with additional data on Ce and Pm).

For Goldschmidt, the time was now ripe for a systematic attack on the problems of crystal structure. Following the discovery of the diffraction of X-rays by crystals by von Laue in 1912, the structure of some simple compounds such as halite (NaCl) and sphalerite (ZnS) had been worked out. These investigations showed that in most inorganic compounds the concept of individual molecules was not valid—a crystal consisted of positively and negatively charged atoms (cations and anions) held together by electrostatic attraction, whose centers were placed at defined points on a geometric lattice. The smallest unit from which this lattice was derived by repetition is known as the unit cell (for example, the unit cell of NaF is a cube with fluorine anions at each corner and the center of each face, and sodium cations at the center of each edge). By 1920 the structure of the atom as postulated by Rutherford and Bohr was generally accepted—a positively charged nucleus surrounded by electrons in orbital shells. It was clear that in crystals the atoms or ions behaved essentially as spheres with a defined radius of the order of 1Å (10^{-8} cm). The size of a unit cell was controlled by the number and sizes of the constituent atoms or ions; however, until the size of a specific ion was known, it was not possible to determine ion sizes from unit cell determinations.

This impasse was broken in 1923 when the Finnish scientist J. A. Wasastjerna determined from optical considerations the ionic radii of fluorine, F^{1-} and oxygen, O^{2-} as 1.33Å and 1.32Å, respectively. Goldschmidt seized upon these data as the key to a complete revolution in the science of crystal chemistry. By determining the unit cell dimensions of oxide and fluoride compounds of the type AX and AX_2 (A being a metal, X oxygen or fluorine) he could determine the ionic radii of the metallic elements;[5] by replacing oxygen or fluorine with other anions, he could then determine anionic radii.

Then followed a period of hectic activity in his Mineralogical Institute and the Raw Materials Laboratory. Between 1924 and 1926 Lunde prepared most of the compounds; Barth made more than 1300 X–ray photographs;[6] and Zachariasen, whose mathematical prowess Goldschmidt greatly admired, determined many of the crystal structures. In May 1926 Goldschmidt was ready with the first table of ionic radii (Table 2). Shortly thereafter, Linus Pauling at the California Institute of Technology, using a different approach through wave mechanics, produced a similar table. The correspondence between the two tables was a brilliant confirmation of the theoretical background.

Goldschmidt summarized the development of his crystal chemistry in the foreword to *Geochemische Verteilungsgesetze der Elemente VII* (1926):

> "This monograph, the seventh in the series *Geochemische Verteilungsgesetze der Elemente,* comprises the theoretical results of crystal chemical studies. The experimental work, which is the basis for the theoretical considerations, was originally undertaken for purely practical purposes, in order to establish the laws of isomorphism, since isomorphic relationships determine in large part the geochemical and technical behavior of the chemical elements. The experimental work in the field of crystal chemistry, which has been carried out during the past few years in the Mineralogical Institute of the university, thus had the purpose of elucidating the geochemical distribution and technical properties of economically important raw materials. These researches, carried out largely for the Norwegian Government's Raw Materials Commission, yielded a series of crystal–chemical generalizations which expanded the work beyond the original considerations. Again and again the experimental work invoked general considerations of crystal chemistry and, I believe, thereby contributed greatly to the original purely practical purposes of our work.
>
> For the geologist and for the chemist, crystallography was previously a purely descriptive science, useful in the recognition and differentiation of crystalline material—either minerals or technical products. Our results have united crystallography and technology on the basis of new principles; crystal chemistry has provided a series of important laws of geochemical association, has enabled predictions of the geochemical distribution of the elements, and

Table 2. Empirical ionic radii (Goldschmidt, 1926) and theoretical ionic radii (Pauling, *J. Amer. Chem. Soc.*, vol. 49, p. 763, 1927), in Ångstrom units.

Atomic number	Element name	Ionic charge	Goldschmidt	Pauling
3	Lithium	1+	0.78	0.60
4	Beryllium	2+	0.34	0.31
5	Boron	3+	—	0.20
6	Carbon	4+	0.2	0.15
7	Nitrogen	5+	0.1–0.2	0.11
8	Oxygen	2−	1.32	1.40
9	Fluorine	1−	1.33	1.36
11	Sodium	1+	0.98	0.95
12	Magnesium	2+	0.78	0.65
13	Aluminum	3+	0.57	0.50
14	Silicon	4+	0.39	0.41
15	Phosphorus	5+	0.3–0.4	0.34
16	Sulfur	2−	1.74	1.84
17	Chlorine	1−	1.81	1.81
19	Potassium	1+	1.33	1.33
20	Calcium	2+	1.06	0.99
21	Scandium	3+	0.39	0.41
22	Titanium	4+	0.64	0.68
23	Vanadium	5+	0.65	0.59
24	Chromium	6+	0.64	0.52
25	Manganese	2+	0.91	0.80
26	Iron	2+	0.83	0.75
27	Cobalt	2+	0.82	0.72
28	Nickel	2+	0.78	0.69
29	Copper	1+	—	0.96
30	Zinc	2+	0.83	0.74
31	Gallium	3+	0.62	0.62
32	Germanium	4+	0.44	0.53
34	Selenium	2−	1.91	1.98
35	Bromine	1−	1.96	1.95
37	Rubidium	1+	1.49	1.48
38	Strontium	2+	1.27	1.13
39	Yttrium	3+	1.06	0.93
40	Zirconium	4+	0.87	0.80
41	Niobium	5+	0.69	0.70
42	Molybdenum	4+	0.68	0.66
44	Ruthenium	4+	0.65	0.63
45	Rhodium	3+	0.68	—
47	Silver	1+	1.13	1.26
48	Cadmium	2+	1.03	0.97
49	Indium	3+	0.92	0.81
50	Tin	4+	0.74	0.71
52	Tellurium	2−	2.11	2.21
53	Iodine	1−	2.20	2.16
55	Cesium	1+	1.65	1.69
56	Barium	2+	1.43	1.35

Table 2. Continued.

Atomic number	Element name	Ionic charge	Goldschmidt	Pauling
57	Lanthanum	3+	1.22	1.15
58	Cerium	3+	1.18	—
59	Praseodymium	3+	1.16	—
60	Neodymium	3+	1.15	—
62	Samarium	3+	1.13	—
63	Europium	3+	1.13	—
64	Gadolinium	3+	1.11	—
65	Terbium	3+	1.09	—
66	Dysprosium	3+	1.07	—
67	Holmium	3+	1.05	—
68	Erbium	3+	1.04	—
69	Thulium	3+	1.04	—
70	Ytterbium	3+	1.00	—
71	Lutetium	3+	0.99	—
74	Tungsten	4+	0.68	0.66
76	Osmium	4+	0.67	0.65
77	Iridium	4+	0.66	0.64
79	Gold	1+	—	1.37
80	Mercury	2+	1.12	1.10
81	Thallium	3+	1.05	0.95
82	Lead	4+	0.84	0.84
90	Thorium	4+	1.10	1.02
92	Uranium	4+	1.05	0.97

above all the new crystal chemistry provides chemists and metallurgists theoretical principles and techniques to supplement the previous empirical procedures" (translated from German).

Goldschmidt established the following rules relating ionic size to atomic structure (Fig. 7):

1. For elements in the same group (vertical column) of the Periodic Table, the ionic radii increase as the atomic numbers of the elements increase. This is, of course, to be expected, since for elements in the same group of the Periodic Table the number of electron orbits around the nucleus, and hence the effective radius, increases in going down the column.
2. For positive ions of the same electronic structure the radii decrease with increasing charge. For an example, we may take the elements in the second horizontal row of the Periodic Table, all of which have two electrons in the inner orbit and eight in the outer orbit: the ionic radii decrease from Na^{1+} = 0.98Å to S^{6+} = 0.34Å. As electrons are lost the nucleus exerts a greater pull on those remaining, thus decreasing the effective radius.

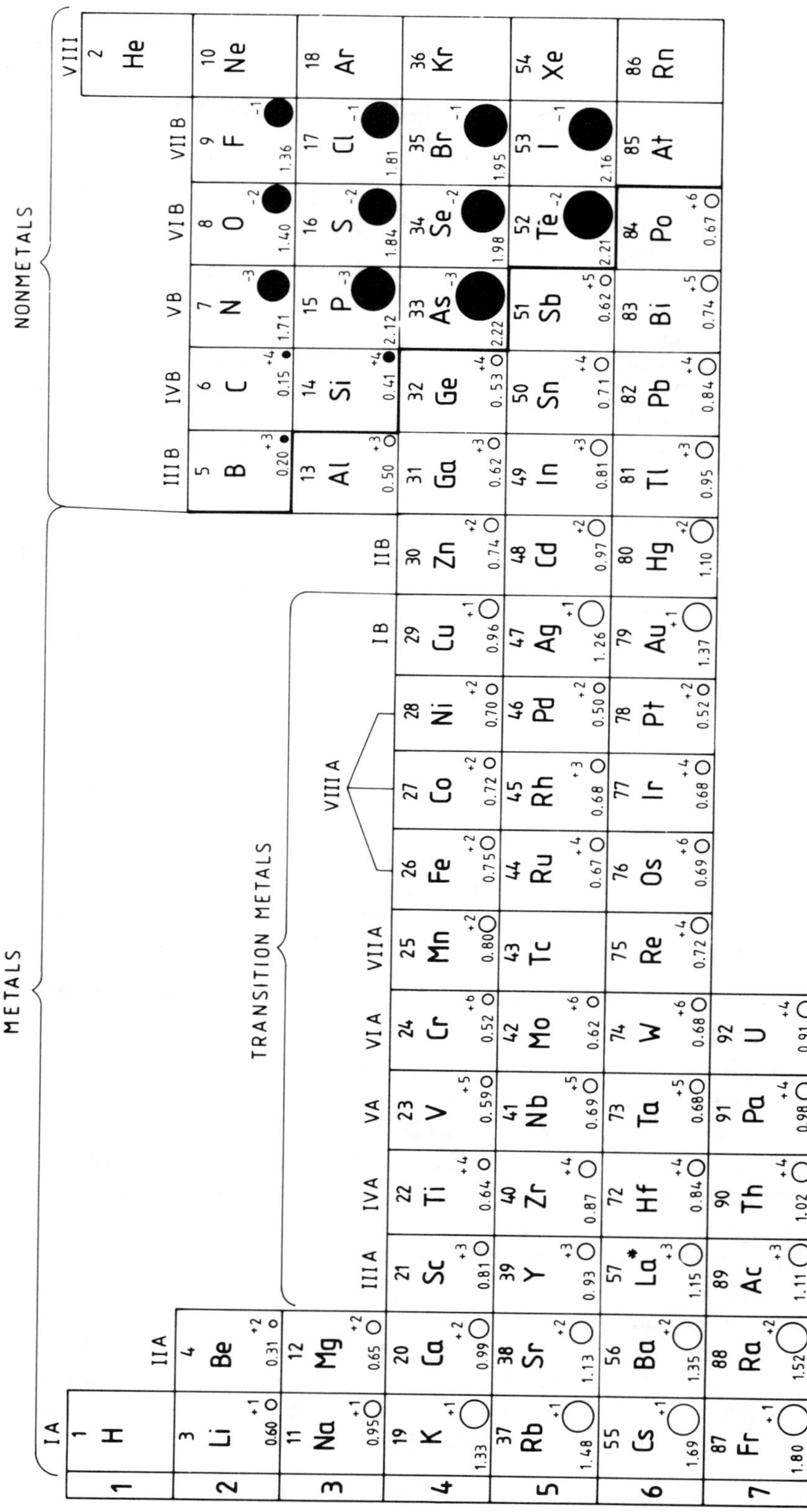
METALS
NONMETALS
TRANSITION METALS
IA
IIA
IIIA
IVA
VA
VIA
VIIA
VIIIA
IB
IIB
IIIB
IVB
VB
VIB
VIIB
VIII
1
2
3
4
5
6
7
1 H
2 He
3 Li +1 0.60
4 Be +2 0.31
5 B +3 0.20
6 C +4 0.15
7 N -3 1.71
8 O -2 1.40
9 F -1 1.36
10 Ne
11 Na +1 0.95
12 Mg +2 0.65
13 Al +3 0.50
14 Si +4 0.41
15 P -3 2.12
16 S -2 1.84
17 Cl -1 1.81
18 Ar
19 K +1 1.33
20 Ca +2 0.99
21 Sc +3 0.81
22 Ti +4 0.64
23 V +5 0.59
24 Cr +6 0.52
25 Mn +2 0.80
26 Fe +2 0.75
27 Co +2 0.72
28 Ni +2 0.70
29 Cu +1 0.96
30 Zn +2 0.74
31 Ga +3 0.62
32 Ge +4 0.53
33 As -3 2.22
34 Se -2 1.98
35 Br -1 1.95
36 Kr
37 Rb +1 1.48
38 Sr +2 1.13
39 Y +3 0.93
40 Zr +4 0.87
41 Nb +5 0.69
42 Mo +6 0.62
43 Tc
44 Ru +4 0.67
45 Rh +3 0.68
46 Pd +2 0.50
47 Ag +1 1.26
48 Cd +2 0.97
49 In +3 0.81
50 Sn +4 0.71
51 Sb +5 0.62
52 Te -2 2.21
53 I -1 2.16
54 Xe
55 Cs +1 1.69
56 Ba +2 1.35
57 La* +3 1.15
72 Hf +4 0.84
73 Ta +5 0.68
74 W +6 0.68
75 Re +4 0.72
76 Os +6 0.69
77 Ir +4 0.68
78 Pt +2 0.52
79 Au +1 1.37
80 Hg +2 1.10
81 Tl +3 0.95
82 Pb +4 0.84
83 Bi +5 0.74
84 Po +6 0.67
85 At
86 Rn
87 Fr +1 1.80
88 Ra +2 1.52
89 Ac +3 1.11
90 Th +4 1.02
91 Pa +4 0.98
92 U +4 0.91

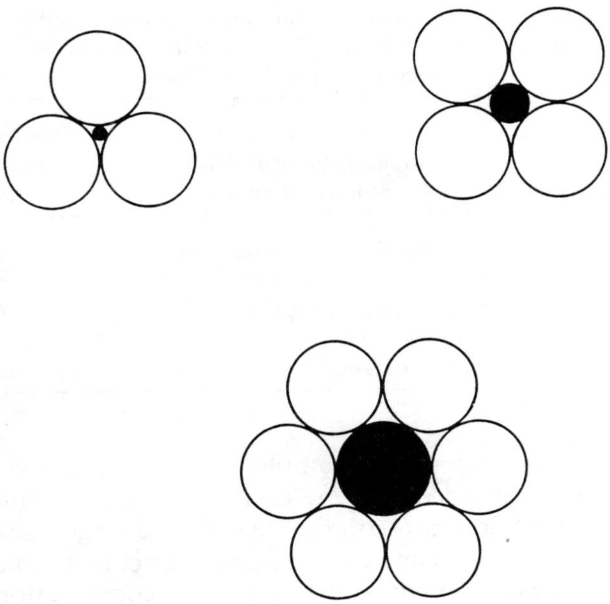

FIG. 8. Planar representation of the relationship between radius ratio and coordination number (after Goldschmidt, *Geochemische Verteilungsgesetze der Elemente VII*, p. 113, 1926).

3. For an element that can exist in several valence states, i.e., form ions of different charge, the higher the positive charge on the ion, the smaller the radius, for example: Mn^{2+} = 0.91Å, Mn^{3+} = 0.70Å, Mn^{4+} = 0.52Å. This is a corollary of the preceding rule.

In an ionic structure (and most inorganic compounds can be considered as ionic structures) each ion tends to surround itself with ions of opposite charge. Goldschmidt demonstrated that the number that can be grouped around the central ion depends upon the radius ratio between the two (Fig. 8). By assuming that ions act as rigid spheres of fixed radii, he calculated the stable arrangements of cations and anions for particular radius ratios from geometric considerations (Table 3). Goldschmidt realized his assumption was not entirely valid; he wrote: "We must regard the edifices of crystal structure not as constructions made from rigid spheres, but as dynamic equilibria, the dimensions (of the atoms and atomic groupings) yielding to some degree to interatomic binding and repulsive forces".[7] Nevertheless, the concept of the radius ratio has proved extremely fruitful in elu-

---

FIG. 7. The Periodic Table of the elements, illustrating ionic sizes; La* occupies the space for the lanthanides, La (57) to Lu (71), their ionic size ranging from La^{3+} = 1.15Å to Lu^{3+} = 0.85Å. Open circles are cations, solid circles anions.

Table 3. Relationship between radius ratio and coordination number for ions as rigid spheres (after Goldschmidt, *Geochemische Verteilungsgesetze der Elemente VIII*, p. 15, 1927).

Radius ratio (R_{cation}/R_{anion})	Arrangement of anions around cation	Coordination number of cation
0.15–0.22	Corners of an equilateral triangle	3
0.22–0.41	Corners of a tetrahedron	4
0.41–0.73	Corners of an octahedron	6
0.73–1	Corners of a cube	8
>1	Closest packing	12

cidating crystal structures. For example, in the many AX compounds with the sodium chloride structure, each cation is surrounded by six anions, i.e., the cation is six-coordinated. With a larger cation–anion ratio, as with the large cesium ion, a different structure becomes stable, the cesium chloride structure, in which the coordination number is eight.

Many cations occur exclusively in a specific coordination; others, which have a radius ratio lying near the theoretical boundary between two types of coordination, may occur in both. Thus calcium can be accommodated in either six- or eight-coordination with oxygen, and aluminum in either six- or four-coordination; sometimes both types of coordination can be present in the same compound.

In his considerations of crystal chemistry Goldschmidt greatly expanded the concepts of *isomorphism* and *polymorphism*, and elucidated many puzzling features of these phenomena. The term *isomorphism* was introduced by the German chemist Mitscherlich in 1819, who prepared crystals of KH_2PO_4, KH_2AsO_4, $(NH_4)H_2PO_4$, and $(NH_4)H_2AsO_4$ and found that their crystal forms were essentially identical. By Mitscherlich's original definition, substances with analogous formulas and similar crystallography were said to be isomorphous. Goldschmidt widened this concept by showing that substances with analogous formulas and with ions of the same coordination usually had the same structure, as demonstrated by the large number of compounds with the sodium chloride structure. Sometimes the chemical analogy was not obvious; Goldschmidt showed that boron nitride, BN, had the same structure as graphite (hexagonal carbon);[8] he speculated that there might be a high–pressure form of boron nitride with the structure and properties of diamond. This cubic form of boron nitride was subsequently synthesized by the General Electric Company in 1957, and is marketed under the trade name of Borazon.

Polymorphism refers to the phenomenon whereby an element or compound crystallizes in more than one structure; for example, diamond and graphite are polymorphic forms of carbon. Goldschmidt showed that many instances of polymorphism are associated with changes in coordination number—in graphite each carbon is linked to three other carbon atoms to form a sheet structure, whereas in diamond each atom is linked to four other atoms in a three-dimensional structure. This explanation for polymorphism and isomorphism was illustrated by Fig. 9. For cations with a radius less than 1.0Å, the borates, nitrates, and carbonates all have the same trigonal crystal structure, that of calcite, $CaCO_3$; for cations with a radius greater than 1.0Å, a different structure (orthorhombic), that of the $CaCO_3$ polymorph aragonite, is the stable form. For divalent calcium, with an ionic radius of 0.98Å, both structures are possible.

Goldschmidt also considered the relationship between crystal structure and other physical properties. He pointed out that, for compounds with the same structure, the hardness was related to the ionic charge. On a scale of 1 to 10 the hardness of compounds with the sodium chloride structure is directly related to the charges on the ions, thus:

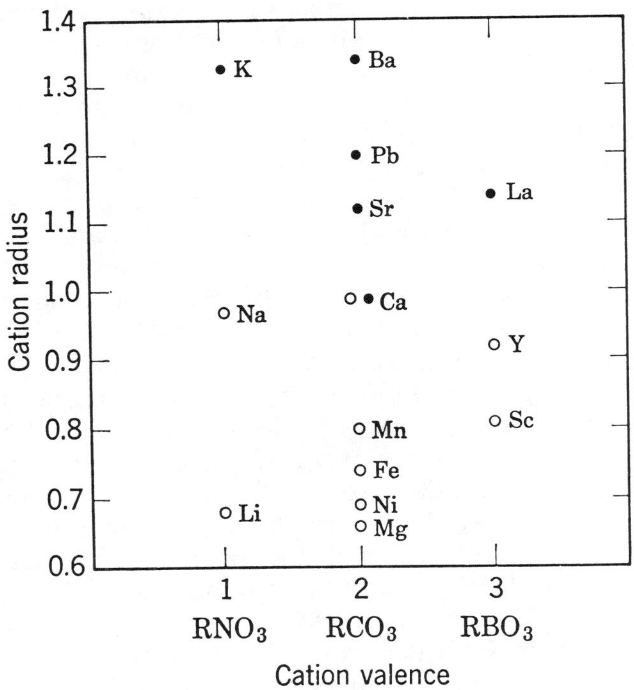

FIG. 9. Effect of cation radius in determining crystal structure: ○ = trigonal; ● = orthorhombic (after Goldschmidt and Hauptmann, *Nachr. Ges. Wiss. Göttingen, Math.-phys. Klasse,* p. 53, 1932).

Compound	NaF	MgO	ScN	TiC
Ionic charge	1	2	3	4
Hardness	3.2	6.5	7 to 8	8 to 9

He developed a *model principle* to construct crystals with a prescribed atomic arrangement and pre–determined physical properties. As an example, he cited Zn_2SiO_4 and Li_2BeF_4, which have the same crystal structure but very different physical properties:

	Zn_2SiO_4	Li_2BeF_4
Hardness	5.5	3.8
Melting Point	1510°C	470°C
Solubility in H_2O	Insoluble	Rather soluble

In subsequent years this principle has been widely applied to synthesize compounds with desirable physical properties.

A great contribution of the new crystal chemistry to the advancement of geochemistry was the prediction of the occurrence and distribution of minor and trace elements, through a knowledge of their atomic and ionic radii. The development of reliable methods for the analysis of minerals showed that many of them are variable in composition; substitution of one element by another is the rule rather than the exception. When this phenomenon was first observed in the mid-nineteenth century it was described in terms of the concept of solid solution or mixed crystals, which implied the presence, in a homogeneous crystal, of molecules of two or more substances. For example, the common mineral olivine, an iron–magnesium silicate whose formula is $(Fe,Mg)_2SiO_4$, was described as a solid solution of Mg_2SiO_4 and Fe_2SiO_4. Since Mg_2SiO_4 and Fe_2SiO_4 are isomorphous, solid solution was often equated with isomorphism, despite the fact that many isomorphous compounds show little or no solid solution, and extensive solid solution may occur between components that are not isomorphous. The new crystal chemistry provided a radically improved interpretation. In an ionic structure there are no molecules, the structure being an infinitely extended network of regularly spaced ions. Any ion in the structure may be replaced by a different ion of similar radius without causing serious distortion of the structure, just as a bricklayer, running short of red bricks, may incorporate yellow bricks of the same size here and there in his wall. Since minerals crystallize from naturally occurring solutions containing many elements other than those essential to the mineral, they often incorporate some foreign ions in the structure. Goldschmidt showed that provided the radii of substituting and substituted ions do not differ by more than 15%, a wide range of substitution may be expected. It is not essential that the substituting ions have the same charge or valency, provided that electrical neutrality is maintained by concomitant substitution elsewhere in the structure. For example, potash feldspar commonly contains some barium: Ba^{2+} substitutes for K^{1+} and electrical neutrality is maintained by the coupled substitution of Al^{3+} for Si^{4+}. These concepts provided the clues that Goldschmidt was to exploit during the next decade in the elucidation of the geochemistry of many rare elements.

In addition to the concept of camouflage, mentioned in the case of zirconium–hafnium, where a rare element is camouflaged in the minerals of a more common element with the same ionic charge and essentially the same size (other examples are Al^{3+}—Ga^{3+}, Si^{4+}—Ge^{4+}); Goldschmidt introduced the concepts of *capture* and *admission*. When a minor element has a similar ionic radius but a higher charge than that of a major element (or the same charge or a lesser radius) it will be preferentially incorporated in a crystallizing mineral, and hence is "captured" by it. Thus Ba^{2+} (1.34Å) is captured by potassium minerals (K^{1+} = 1.33Å). On the other hand, when a minor element has a similar ionic radius but a lower charge than those of a major element, it is less readily incorporated, and is said to be "admitted" into those minerals containing the major element. Thus Li^{1+} (0.78Å) is admitted into magnesium (Mg^{2+} = 0.78Å) minerals. In capture and admission of ions of different charge the charge balance is maintained by concomitant substitution elsewhere in the crystal lattice.

After 1926 the tempo of research at the Mineralogical Institute slowed somewhat. Ulrich had returned to Prague. Thomassen had left for the California Institute of Technology in December 1924; he returned in October 1927 but resigned in July 1929 to accept a professorship in chemical engineering at the University of Michigan. Barth took his doctoral examination in September 1927 and accepted a position at the Technical University of Berlin; in this connection, Goldschmidt wrote the following recommendation: "Since autumn 1924 he has been scientific assistant at the Mineralogical Institute. I praise his personal qualities, as an extremely reliable, dutiful, and diligent co-worker." Zachariasen produced a 165-page monograph on the structure of ABO_3 compounds for his doctoral examination in October 1928 and went to do research with Sir Lawrence Bragg in Manchester, and from there to a professorship in the University of Chicago in 1930. In 1929 Lunde became director of the Norwegian Cannery Association laboratory in Stavanger. After 1927 Goldschmidt was preoccupied with the proposed move to Göttingen.

In 1927, however, Goldschmidt acquired an outstanding co-worker, Felix Machatschki,[9] from Graz in Austria, who came to the Mineralogical Institute with a Rockefeller Scholarship. He worked with Goldschmidt for some years, first in Oslo and then as guest lecturer in Göttingen in 1929–30, except for a period with Sir Lawrence Bragg in Manchester in 1928–29. He absorbed Goldschmidt's ideas on crystal chemistry and isomorphism and immediately applied them to significant mineralogical problems. He attacked the problem of silicate structures, which Goldschmidt had largely avoided because of their complexity. In a short paper published early in 1928 on the structure of the feldspars he demolished current theories of silicate structure based on hypothetical silicic acids, and established the concepts on which a new theory was built. As the fundamental unit of silicate structures he proposed a tetrahedral group of four O^{2-} ions around a central Si^{4+} ion. This Si^{4+} ion could be replaced by an Al^{3+} ion, also in four–coordination with oxygen, but aluminum could also occur in six–coordination with oxygen, a dual role which was to explain many otherwise puzzling mineral compositions. The tetrahedra could be linked through oxygen atoms to form a variety of structures. In the feldspars every oxygen ion is shared by two adjacent tetrahedra and the ratio Si + Al:O = 1:2, giving the formulae $KAlSi_3O_8$ (orthoclase), $NaAlSi_3O_8$ (albite), and $CaAl_2Si_2O_8$ (anorthite) and explaining the complete solid solution between albite and anorthite (the plagioclase feldspars), despite the difference in their formulas.

In 1929 Machatschki secured another triumph when he derived a satisfactory formula for tourmaline, whose chemistry John Ruskin described in 1891 as "more like a medieval doctor's prescription than the making of a respectable mineral." By a careful evaluation of many chemical analyses, and the application of the new concepts of crystal chemistry, he derived the formula $XY_9B_3Si_6H_xO_{31}$; in X and Y he grouped cations of similar size, as follows: X = Na^{1+}, Ca^{2+}; Y = Li^{1+}, Mg^{2+}, Mn^{2+}, Fe^{2+}, Al^{3+}. He showed that the unit cell of tourmaline contained three of these formula units.

Other visiting scientists from abroad were Dorothy Wyckoff,[10] a petrologist from Bryn Mawr College in Philadelphia, and Adolf Pabst, a mineralogist from the University of California, who spent the academic year 1928–29 at the Mineralogical Institute. Goldschmidt wrote of Pabst (letter, March 14, 1929): "He must be classed among the very best scientists who have visited my laboratory, certainly among the best five". Pabst went on to a distinguished career as Professor of Mineralogy at the University of California at Berkeley, and received the premier award of the Mineralogical Society of America, the Roebling Medal, in 1965.

In 1927 Goldschmidt's mother suffered a stroke with paralysis on one side, and the Goldschmidt household employed a housekeeper, Miss Marie Brendingen, who remained a devoted servant until Goldschmidt's death in 1947. During 1928 he was frequently absent, giving invited lectures in northern Europe and supervising the construction of a new Mineralogical Institute in the University of Göttingen. During that summer he developed a serious lung condition as a result of the incautious use of hydrofluoric acid in his experiments; he lost his voice completely for some time, and spent the month of July at the spa at Bad Ems recuperating. While there he wrote to Zachariasen (July 24, 1928) " . . . Yesterday and the day before I was visited by Dr. Bernal from Cambridge. We talked crystal chemistry for up to 12 hours, without me becoming hoarse. . . . " Bernal had a distinguished career, both in science and politics; he gave the memorial address to the Chemical Society of London after Goldschmidt's death.

Goldschmidt was nominated for the Nobel Prize in chemistry for 1929.[11] However, the prize for that year went to Arthur Harden and Hans von Euler, for their researches on fermentation processes. It took some years for the significance of the new crystal chemistry to be recognized in wider scientific circles; I can recall no mention of it in my graduate studies in chemistry in the University of New Zealand in the late 1930's.

In March 1929 Goldschmidt gave two invited lectures in London. One, to the Faraday Society, entitled *Crystal Structure and Chemical Composition,* was a comprehensive exposition of the new crystal chemistry, as developed by him and his associates in Oslo. The other, to the Royal Institution, was entitled *The Distribution of the Chemical Elements,* and outlined the development of his ideas on geochemistry since the publication of his *Der Stoffwechsel der Erde* in 1922. An example in this paper illustrates his genius in obtaining a simple solution to a complex problem, in this case the average composition of the Earth's crust. He noted that if it were possible to obtain an average sample of a large part of the Earth's crust consisting mainly of crystalline rocks, its chemical analysis would give a reliable picture of the composition of the crust as a whole. Such an average sample was provided, he pointed out, by the glacial clay widely distributed in southern Norway, which represents the finest rock flour deposited by melt water from the Fennoscandian ice sheet. During previous years the Raw Materials Laboratory had made numerous analyses of these clays for technological purposes. From 81 analyses of different samples of this material he calculated the following average:

SiO_2	Al_2O_3	$FeO+Fe_2O_3$	MgO	CaO	Na_2O	K_2O	TiO_2	P_2O_5	H_2O
59.19	15.68	6.96	3.38	3.20	2.10	3.74	0.80	0.21	3.06

In 1924 F. W. Clarke and H. S. Washington in the U.S. had derived an average composition of the igneous rocks of the Earth's crust, based on Washington's collection of 5159 superior analyses selected from the literature. Their results, calculated to 100 percent with the elimination of H_2O, are as follows:

SiO_2	Al_2O_3	Fe_2O_3	FeO	MgO	CaO	Na_2O	K_2O	TiO_2	P_2O_5
60.18	15.61	3.14	3.88	3.56	5.17	3.91	3.19	0.16	0.30

As Goldschmidt commented, the correspondence between the two sets of figures is remarkable, especially when the effects of hydration and solution, which result in the leaching of sodium and calcium from the clays, are taken into account.

Goldschmidt also applied his data on ionic radii to an intriguing consideration of the composition of the Earth's crust or lithosphere. The Earth's crust consists almost entirely of oxygen compounds, mainly silicates of aluminum, calcium, magnesium, sodium, potassium, and iron. Nine elements—O, Si, Al, Fe, Ca, Na, K, Mg, and Ti—make up over 99% of the total. Of these, oxygen is absolutely predominant. As he pointed out, this predominance is even more marked when the figures are recalculated to atom percent and volume percent (Table 4). In terms of number of atoms, oxygen exceeds 60%. If the volume of the different atoms, or rather ions, is calculated, oxygen makes up more than 90% of the total volume occupied by the elements. Thus the Earth's crust is essentially a packing of oxygen anions, bonded by silicon and the ions of the common metals. As Goldschmidt remarked, the lithosphere may well be called the oxysphere.

Table 4. The composition of the Earth's crust, in terms of weight percent, atom percent, and volume percent of the elements (Goldschmidt, *Neues Jahrb. Mineral. Beilage, Abt. A*, vol. 57, p. 1119, 1928).

Element	Weight %	Atom %	Radius (Å)	Volume %
O	46.59	62.46	1.32	91.77
Si	27.72	21.01	0.39	0.80
Al	8.13	6.44	0.57	0.76
Fe	5.01	1.93	0.82	0.68
Mg	2.09	1.84	0.78	0.56
Ca	3.68	1.93	1.06	1.48
Na	2.85	2.66	0.98	1.60
K	2.60	1.43	1.33	2.14
Ti	0.63	0.28	0.64	0.22

On Goldschmidt's departure from Norway in 1929, King Haakon honored him by making him a Knight (First Class) of the Royal St. Olav's Order, for services to Norwegian science.

On September 26, 1929, Goldschmidt gave an invited lecture at the University of Helsinki. After the lecture his long–time colleague and friend, Pentti Eskola, spoke as follows:[12]

> "Honored guest, Professor Goldschmidt, it is for me a pleasant duty to thank you for your visit and for sharing with us the results of your researches. I recall a conversation with you a long time ago as to the value of lectures, now that scientific results are so readily communicated by publication. You replied that personal communication could transmit ideas and inspiration to the listeners far better than the printed word. Your lecture today has demonstrated the truth of that reply, and is a stimulus to our scientific work.
>
> Since I had the good fortune some years ago to work closely with you on a daily basis, and thus know you better, perhaps, than most of your audience, I would like to speak about Professor Goldschmidt rather than to him. I want to express my great admiration for him, both for his impressive personality and for his enormous

scientific achievements. There is something titanic about him, including his discovery of titanium dioxide as a pigment, of which your audience may be unaware. I can truly say that he has fought a titanic fight against many difficulties, not least health problems, which many times have threatened his scientific career. It is only his remarkable energy and his ability to work both day and night that has enabled him to achieve his unique scientific results. As he himself has told me, genius is nothing more than extreme perseverance. I think that is an exaggeration, inspiration is also needed, but perseverance is certainly an important factor.

When, at the age of twenty–three, he published his great work *Die Kontaktmetamorphose im Kristianiagebeit* petrographers throughout the world at first had difficulty in appreciating his concepts. Even if it was realized that petrology is essentially physical chemistry applied to rocks, this was considered of purely theoretical significance. Many scientists were astounded by this demonstration of the practical applications. Younger researchers, not too indoctrinated in current petrographic theory, were able to apply his new principles. I was one of them, and ever since have endeavored to apply chemical theory to the study of rocks. This led to collaboration between Goldschmidt and myself, which ten years ago took me to Oslo, where I stayed for almost a year.

This was a year of intensive work for me. I often discovered how my ideas developed through conversations and discussions with Professor Goldschmidt. I will give one example. One morning, when I was on my way up to my room on the fourth floor of the Mineralogical Institute, I met Professor Goldschmidt on the stairs between the second and third floors. At that time I was studying the nature and origin of eclogites. We began to talk about eclogites and the interior of the Earth. We stood there several hours, until we noticed other staff members leaving for lunch! We had talked the morning away, but that morning was most stimulating for me.

Ever since 1906, when the 18–year Goldschmidt developed a method for measuring the radioactivity of minerals, hardly a year has gone by without his making some significant scientific discovery. It would take too long to enumerate them. During my time in Oslo petrology was his primary concern, although he had already begun his researches into crystal chemistry. In his lectures he often expressed completely original ideas, and when I listened to this youthful professor I frequently had the impression that this was a wise old man deciphering nature's secrets. There was often a certain personification of atoms and crystals, that their secrets could only be unlocked through some sort of cunning. I think I discerned something of this approach in the lecture we have just heard. In his lectures during my Oslo visit he discussed the ideas which were not published until some years later in the classic papers *Stammestypen der Eruptivesteine* and *Der Stoffwechsel der Erde.* Some of his ideas were published in a preliminary form in his monographs on the Stavanger region, which were especially significant for us in Finland, because he expounded on the role of metasomatism in the formation of migmatites. When Professor Goldschmidt lectured on silicate metasomatism in Stockholm, a Swedish geologist, Harald Johansson, commented that the lecturer was accusing the chemical elements, previously known for stable occurrence, for a type of loose living. Goldschmidt replied that this was not unknown even in the best families! It is now realized, thanks largely to Professor Goldschmidt's researches, that metasomatism is a very important process in the formation and alteration of rocks.

It should be mentioned that Professor Goldschmidt completed the Stavanger manuscript in an afternoon and evening [1919] while he was suffering from peritonitis. He wanted to complete the manuscript because he was doubtful of surviving his illness. When late in the evening the manuscript was finished he went directly to the hospital and was immediately operated on.

Thereafter he began the researches which are his greatest achievement till now, whose results have been published in a series of eight monographs entitled *Geochemische Verteilungsgesetze der Elemente.* Professor Wasastjerna is more competent than I to comment on the crystal chemistry. However, in these monographs Professor Goldschmidt has founded a new scientific field more closely linking geology with physics and chemistry than ever before. We must wish him all success in his continued work in this field, but as a petrologist I hope that he will continue to consider himself as a petrologist and mineralogist and not only as a chemist" (translated from Swedish).

7 / *Geochemistry in Göttingen: 1929–1935*

DURING THE 1920'S GOLDSCHMIDT received several invitations to professorial posts in European universities. On May 18, 1924 Richard Willstätter, professor of chemistry at the University of Munich (who received the Nobel Prize in Chemistry in 1915 for his work on plant pigments) wrote to him asking whether he would consider being the successor to Paul von Groth as professor of mineralogy. Willstätter describes what happened in his book *From my Life:*[1]

> "In the summer of 1924 my old teacher and admired colleague, Paul von Groth, retired from his professorship at the age of eighty–one because of long–standing infirmities of old age which he was just beginning to notice himself. . . . He had already explained to us many years earlier that there was only one man whom he wanted and considered suitable as his successor, though not a product of his school—Victor M. Goldschmidt, who was a son of the respected chemistry professor Heinrich Goldschmidt. . . . Victor M. Goldschmidt was master of scientific fields no less wide in scope than Groth's, but in different fashion. His gift lay in profound searching by productive experimental work, and he was able to open up such great new fields as modern geochemistry by his widely recognized, trail–blazing researches. Wien (professor of physics and dean of the science faculty) opposed Goldschmidt's appointment, if not before the assembled faculty, at least individually with most of our colleagues. Eventually he was seen going from place to place in the Academy, preparing the faculty vote. He was dean now and knew how to conduct the discussion along his own lines. Wien would never mention Jews or non–Aryans, but he talked of foreigners. Goldschmidt's appointment was rejected, although the faculty was unable to make another similarly appropriate nomination. The Ministry appointed a professor from a girl's high school whom the faculty had not nominated and did not approve of. Groth was profoundly unhappy.
>
> If, in the given circumstances, the faculty was disinclined by a large majority to nominate the only famous scholar who had been

> nominated, it was because they did not want a scholar of presumably Jewish descent (or probably partly Jewish, I know nothing about it) in their midst. I announced my resignation on the evening after this vote."

It may be recalled that Munich was the cradle of Naziism, where the party was formed after World War I. Goldschmidt evidently had some misgivings, even before the faculty vote. He wrote to Willstätter ". . . Racial fanaticism is one of the evil phenomena of the present day, and I fear it will spread over the whole world. . . . In the spring of 1914 the choice between Munich and Kristiania would not be in doubt; today I have serious reservations about Munich; no one knows what will be the state of the world in 1934" (translation from German). A truly prescient statement! Willstätter, himself a Jew, was threatened with arrest by the Gestapo in 1938 but he escaped to Switzerland, where he died in 1942.

In the summer of 1925 Goldschmidt received an extremely attractive offer from the Technical University in Berlin. He was promised a minimum salary of 28,000 marks, 80,000 marks for equipment of his institute, four assistants, and an additional professor to relieve him of a large part of the teaching duties. This represented a great improvement on his situation in Oslo, where, in addition, he was increasingly isolated and at odds with his colleagues. His scientific renown, and his success in developing his industrial interests, did not sit well with some of his colleagues in the university community. He was not the most diplomatic of men, and he could denigrate a colleague by a thoughtless comment. Goldschmidt accepted the Berlin offer, but pressure from his mentor and friend Professor Brøgger, and from Gunnar Knudsen, the former Norwegian prime minister, persuaded him to withdraw his acceptance and to remain in Norway.

Goldschmidt's dissatisfaction with his situation in Norway is expressed in a letter to Brøgger dated July 20, 1925:

> "When I now, after 16 years of service to the university, find every reason to accept the German invitation, my motivation lies not only in the poor material situation here, but to a large part in the atmosphere of narrow–mindness, envy, stupidity, and malice among some colleagues, which poisons the scientific life. In such an atmosphere one loses hope at achieving any significant contribution to science or the nation" (translated from Norwegian).

Further offers were not long in coming. Otto Mügge, Professor of Mineralogy at Göttingen since 1908, was to retire in 1928. As a result of an invitation, Goldschmidt wrote to the Rector of the University of Oslo on August 29, 1927, requesting leave from September 3–15 "to negotiate the conditions for assuming the mineralogy professorship in Göttingen." The negotiations were to continue over the next two years. He spent January and February 1928 as guest lecturer in Göttingen.

The University of Göttingen was anxious to recruit Goldschmidt, and the Prussian Ministry of Education offered him facilities far superior to those he had in Oslo. He was to have a Mineralogical Institute built to his specifications, with appropriate staffing of research and technical assistants and adequate financing. The university acquired a former secondary school building on Lotzestrasse and equipped it according to his requirements. Goldschmidt described it as follows:[2]

> "When I was invited to Göttingen, the university desired that I should organize a large new mineralogical institute, equipped for crystal–chemical and especially for the geochemical researches which I wished to undertake. Göttingen also offered a great advantage, in that a group of outstanding physicists was interested in collaborative work. I also had the opportunity to promote investigations of industrial importance, such as on refractories for metallurgical use, a subject I was then engaged in.
>
> In Göttingen two buildings were placed at my disposal, a secondary school with an adjoining gymnasium and chemical laboratory. These I had connected with a hall in which the mineral collections were housed [Fig 1 in the original; Plate 17 here]. In the main building the cellar was converted into workshops, a mechanical shop with places for five workers, a carpenters' shop for two men, a laboratory for the preparation of rock and mineral sections, and a high temperature–high pressure laboratory. The library and staff offices were on the first floor; the chemistry section was installed on the second floor, with six laboratories and a balance room; the third floor comprised the X–ray laboratories for crystal structure and chemical analysis. The adjoining building housed lecture rooms and laboratories for teaching mineralogy, petrography, and crystallography, and rooms for research students and visitors. The institute was thus a combined institute for mineralogy, chemistry, and physics, since all three sciences are an essential part of modern geochemical research. In addition we had a small separate building as a rock store and for cutting and grinding. Much of the equipment for teaching and research was made in our own workshops.
>
> The room for X–ray spectrography is illustrated in Fig. 2 [not available]. Some of the apparatus I brought from Oslo and it was installed by my assistant K. Stenvik, who accompanied me the first years in Göttingen. X–ray spectrography is the ideal technique for the analysis of most elements in concentrations greater than 0.1–0.01%, and can be used for lower concentrations if suitable enrichment procedures are available. As examples I mention the work of my Japanese collaborator Dr. E. Minami on the determination of the concentrations of rare earth elements in shales, the first time these elements have been measured in a common rock, and Dr. W. Noll's investigation of the geochemistry of strontium.
>
> The research laboratory for emission spectrography is illustrated in [Fig. 3 in the original; Plate 18 here]. We see three of my research

associates, Dr. H. Hörmann, Dr. H. Bauer, and Dr. L. W. Strock at the large spectrographs specially constructed for our work by Dr. R. Mannkopff. For our researches we required spectrographs with very great dispersion, in order to make photometric analyses of individual lines in the spectrograms" (translated from Norwegian).

After Mügge's retirement and before Goldschmidt's arrival the day–to–day operations of the institute were the responsibility of docent (assistant professor) Dr. Friedrich Heide.[3] In March 1929 Theodor Ernst[4] joined the staff of the institute and worked closely with Goldschmidt during his time in Göttingen.

Goldschmidt's father retired as Professor of Chemistry at the University of Oslo at the end of the 1928–29 academic year. There were two candidates for the vacant chair: Ellen Gleditsch,[5] an accomplished radiochemist, and Odd Hassel,[6] who had worked with Goldschmidt on problems of crystal chemistry. Goldschmidt strongly supported Hassel for the position, but the faculty voted to appoint Gleditsch. For Goldschmidt this was the last straw; he immediately resigned his professorship and accepted the Göttingen post. Odd Hassel was appointed Professor of Physical Chemistry at Oslo in 1935 and received the Nobel Prize in Physics in 1969.

Goldschmidt, his father and mother, and their housekeeper, Miss Brendingen, moved to Göttingen in September 1929, to a large Victorian house at Wagnerstrasse 8.[7] His mother died on October 10, 1929.

In 1929 the science faculty of the University of Göttingen was an unparalleled galaxy of talent. The mathematical tradition was carried on by Hilbert, Courant, Landau, and others. The organic chemist Wallach (Nobel Prize, 1910) was living in retirement, and his successor Windaus (Nobel Prize, 1928) was at the height of his career; his colleague in chemistry, Zsigmondy, had received the Nobel Prize in 1925. The Physics Department was guided by the triumvirate James Franck (Nobel Prize, 1925), Max Born (Nobel Prize, 1954), and R. Pohl. Prandtl, one of the founders of modern aerodynamics, was one of Göttingen's great men, as was Tammann, the physical chemist who was still active in retirement. The zoologist Kühn and the astronomer Kienle became great personal friends of the Goldschmidts. This galaxy of talent, alas, was to be cast to the winds after the Nazi takeover in 1933.

After Goldschmidt's death, Max Born wrote as follows:[8]

> "When he first came to Göttingen there was a latent tension between us, a mutual doubt regarding our work on ionic crystals: he considered my calculations of lattice energies cumbersome and scarcely worthwhile, and I had no confidence in his primitive way of adding ionic radii. But we soon discovered that these two methods were complementary, one providing exact values in a few simple cases, the other covering the whole field of crystal chemistry. After clearing the air, we became great friends. He was a remarkable

> man in many respects. His memory was stupendous. Once I sent an assistant to him to ask about literature on a great number of data. Instead of giving titles of papers, he just dictated the numerical tables from memory, and a later check proved them essentially correct. He knew by heart not only the properties of atoms and ions, like radii, polarizabilities, etc., but also their occurrence in minerals and the distribution of these minerals in the earth's crust. If he did not know a fact, he had simple ways of deriving it from his radii.
>
> Goldschmidt had a very original mind and a great sense of humour which sometimes found expression in a somewhat caustic manner. One had to take the trouble to look below the surface to discover a gentle and most generous nature which was easily wounded, and had no other defence than to hit out. To those who came to understand this, he was a devoted and most loyal friend, and they now mourn for one who had enriched their lives."

Goldschmidt brought some equipment which was his personal property with him to Göttingen, and his personal assistants Kristoffer Stenvik and Mimi Johnson–Høst to help in setting up the laboratories. Felix Machatschki accompanied him as guest lecturer in mineralogy in 1929, leaving in 1930 to become Professor of Mineralogy at the University of Tübingen.

Goldschmidt came to Göttingen in 1929 with great plans for comprehensive investigations on the abundance and distribution of all the chemical elements. His crystal chemistry had elucidated the probable mode of occurrence of the rare elements. In the crystallization of igneous rocks, the rare elements would tend to occur in minerals which would accommodate them in terms of their ionic radii and coordination numbers—in effect, the crystal structures acted as a sorting mechanism, accepting those elements which could be accommodated in specific lattice positions, and rejecting those either too large or too small. For example, nickel is present in minor amounts (up to about 0.4%) in olivine, $(Mg,Fe)_2SiO_4$, because Ni^{2+} has practically the same radius as Mg^{2+} and Fe^{2+}. Elements with ions either too small or too large to be accommodated in the common minerals of igneous rocks concentrated in the residual liquids from magmatic crystallization, which formed pegmatites and quartz veins where the rarer elements were often present as specific minerals.

It was clear to Goldschmidt that classical methods of chemical analysis were inadequate for determining the low concentrations of the rarer elements in rocks and minerals. Even X–ray spectrography, which he had used so successfully in his investigation of rare earth minerals, had a lower level of sensitivity around 0.01% (100 ppm—parts per million). He needed a technique a hundred times more sensitive, down to 1 ppm. He found it in the development of optical spectrography with the carbon arc, a technique known for many years but which had not been applied to the quantitative analysis of rocks and minerals. He recruited a chemist and a physicist, Clemens Peters[9] and Reinhold Mannkopff,[10] and a talented mechanic and instrument maker, Hermann Albrecht. Peters and Mannkopff refined the technique of quantitative spectral analysis with the carbon arc, and Albrecht built and maintained the equipment. The first application, in 1930, resulted in the discovery of germanium as a common constituent of coal and of the soot and ashes therefrom. A story has it that Miss Brendingen found Goldschmidt on his back in front of the fireplace raking soot out of the chimney, and feared that the professor had gone mad! He found that raw coal could contain up to 0.01% Ge, and the soot and ashes could contain 0.1% or more. One ash, from the Hartley seam in the Newcastle district in England, contained 1.6% GeO_2, and was later utilized as a commercial source of this element.

When Goldschmidt arrived in Göttingen there were few students in the Mineralogical Institute, because of the long interregnum since Mügge's retirement. His reputation soon attracted talented students and researchers, not only from Germany but also from abroad. Among the distinguished foreign researchers were Harry Berman[11] from the United States (September, 1932–April, 1933), Mosè Balconi from Italy (October, 1932–December, 1933), Vladimir Shcherbina from Leningrad (September, 1932–March, 1933), Eiiti Minami[12] from Tokyo (May, 1933–July, 1935), Lester Strock[13] from Philadelphia (May, 1933–November, 1935), and others such as Thure Sahama from Finland who visited for shorter periods.

During the Göttingen years the geochemistry of many elements was intensively investigated, as follows:

Element	*Investigators*
Gallium	V. M. Goldschmidt and C. Peters
Scandium	V. M. Goldschmidt and C. Peters
Beryllium	V. M. Goldschmidt and C. Peters
Noble Metals (Ru, Rh, Pd, Ag, Os, Ir, Pt, Au)	V. M. Goldschmidt and C. Peters
Boron	V. M. Goldschmidt and C. Peters
Germanium	V. M. Goldschmidt and C. Peters
Arsenic	V. M. Goldschmidt and C. Peters
Alkali Metals (Li, Rb, Cs)	V. M. Goldschmidt, H. Berman, H. Hauptmann, C. Peters, H. Bauer, H. Witte[14]
Lithium	L. Strock
Strontium	W. Noll[15]
Barium	W. von Engelhardt[16]
Rare Earths (Y, La-Lu)	E. Minami
Chromium	V. M. Goldschmidt, H. Bauer, H. Hörmann

Molybdenum	M. Balconi, H. Hauptmann
Selenium	V. M. Goldschmidt, O. Hefter, L. Strock, E. Minami

The years in Göttingen, up to 1933, were probably the happiest in Goldschmidt's life. In Oslo he was something of a loner, and had become progressively alienated from many of his university colleagues. In Göttingen he not only inspired his own students but drew inspiration from his colleagues—the physical chemists, inorganic chemists, physicists, astronomers, and biologists. It was an almost ideal cooperation with his fellow scientists, and in addition there was a constant exchange of ideas with foreign visitors and guests.

Among the distinguished visitors to Goldschmidt's insitute during the Göttingen years was the noted Russian geochemist V. I. Vernadsky,[17] who at that time was living in Prague. They had known each other for many years (Vernadsky had been a visiting lecturer in Olso in the 1920's), and corresponded frequently.[18] At Goldschmidt's invitation Vernadsky spent several days in Göttingen in June 1932 and gave two lectures. Goldschmidt's letter to him (April 29, 1932) regrets that it is not possible to offer an honorarium, but asks him to be his house guest. Vernadsky arrived on June 15 and Goldschmidt met him at the railroad station. The meeting was an enjoyable one (Plate 23). After Vernadsky's return to Prague, Goldschmidt wrote (July 16)

> "For my father and I your visit to Göttingen was a most pleasant experience, and for my colleagues and students your lectures were of the greatest interest and value. We greatly appreciate your visit, and hope you will return often to Göttingen."

He then goes on to educate Vernadsky on the new developments in crystal chemistry, specifically on the relationship between zoisite, $Ca_2Al_3(SiO_4)_3(OH)$, and orthite (allanite), a complex rare–earth bearing silicate mineral. Already in 1911 Goldschmidt had shown that the two minerals were closely related, probably isomorphous. At that time it would have been assumed that the trivalent rare earths replaced trivalent aluminum. The new crystal chemistry developed by Goldschmidt in the 1920's showed that this explanation was improbable, since the ionic radii of the rare earth elements are much greater than that of aluminum. In 1930 Machatschki, working in Goldschmidt's institute, provided the correct explanation—the rare earths replace calcium, and electrical balance is maintained by the concomitant replacement of some aluminum by magnesium, i.e., $Ca^{2+}Al^{3+}$ by $La^{3+}Mg^{2+}$. Vernadsky was evidently unaware of these developments, and argued that zoisite and orthite could not be isomorphous because orthite is soluble in acids whereas zoisite is not. Goldschmidt replied by pointing out that, in the isomorphous series of plagioclase feldspars, albite ($NaAlSi_3O_8$) is insoluble, whereas anorthite ($CaAl_2Si_2O_8$) is soluble in acids. The correspondence on this subject continued for some time; it is not clear whether Vernadsky was ultimately convinced.

Th. Ernst, assistant to Goldschmidt during the Göttingen years, later Professor of Mineralogy at the University of Erlangen, recalls that time:[19]

> "How did Goldschmidt organize his day? Before coming to the institute he dictated his more private mail to his secretary at his home in Wagnerstrasse, a large old–fashioned villa where he lived with his father. When he arrived at the institute he would attend to administrative matters, and then (if no lectures or laboratory work were scheduled) he would visit his coworkers. These almost daily rounds showed that the "Chief" was always available to discuss the research work and the researcher's problems. If one had a problem, a discussion with Goldschmidt frequently provided a new perspective. He welcomed ideas from his coworkers, while examining them critically.
>
> In discussions one could only admire his remarkable memory, not only for the literature (for which he could often give volume number and year) but also for crystal data. If shown a mineral, he would recall where it was described, to the amazement of coworkers and visitors."
>
> Goldschmidt was an inspired teacher. From the beginning he would be present during the practical work with the petrographic microscope. Naturally, the small number of advanced students learned much from him, but his assistants (Laves[20] and I) learned even more. He prepared his lectures with great care. I was his assistant for some years, and it was not always easy to fill his requests. At first there was almost nothing to illustrate crystal structures. I spent many night hours making models of the principal structure types by glueing together celluloid balls of different sizes, which taught me a lot about the crystalline state. His lectures, so stimulating for the serious student, were delivered in a high–pitched monotone which induced somolence in some; however, he would occasionally animate his listeners with an appropriate joke.
>
> If his instructions were not followed or minor mishaps occurred, as can happen in any institution, he could become very angry and "blow up." Sometimes he recognized that he had overreacted, and then he would make a diplomatic retreat by the gift of a bottle of schnaps or a packet of cigars. With the women in the institute he was very restrained with his appreciation. I remember once when he appeared several times in the chemistry laboratory with a small packet of candy, but did not hand it to the lady in question until the male staff were no longer present.
>
> As Director, Goldschmidt typified the conscientious civil servant, who saw to it that the budget was prudently and properly administered. Laves and I were largely responsible for implementing administrative directives, which occasionally raised some problems with visiting scientists.
>
> With his academic colleagues Goldschmidt—perhaps because of his conservative political leanings—was closest to the experi-

mental physicist R. Pohl and the geneticist R. Kühn. They often accompanied him on short excursions, when Goldschmidt was at his most relaxed (Plate 28). His relationship with some of his Jewish colleagues was somewhat tense, probably because of differing political views. However, it is typical for his upright attitude that in 1933 he publicly affirmed his Jewish faith.

On his departure from Göttingen in the autumn of 1935, his closest co–workers gathered at the railroad station to say goodbye to him and his father. In spite of his difficult situation, Goldschmidt had retained his position with the justification that he had been entrusted with the institution and should remain as long as the higher administration wanted him. He would probably have been spared much grief, if he had left this now unfriendly land much earlier. We remember him with gratitude. As his students we are proud of our teacher; we appreciate his humanity and regret his comparatively early death" (translated from German).

Goldschmidt was a great animal lover and Paul Rosbaud recalls the situation in Göttingen:[21]

> "The household, in addition to its human inhabitants, consisted of the famous dachshund "Bazi"—which tyrannized father and son; three squirrels demanded and received a home in the bathroom, a toad hibernated in the cellar, and a family of bats resided in the loft. Each had a name, and the toad shared his with a well–known and not too endearing character in Göttingen. One of the squirrels was named Parsifal ("because he was incredibly stupid") another was Richard[22] ("Weil er so stark riecht"), but the most notorious was Magdalena, who came to a sad and untimely end. V.M. [Goldschmidt] gave a birthday party for Bazi and several dog guests were invited to share the birthday cake. In the excitement, Bazi concluded that on this special day everything was permitted and ended the feast by devouring Magdalena."

During the Göttingen years Goldschmidt, while directly concerned with the abundance and distribution of the individual elements, was also pondering the wider implications in terms of the geochemical cycle (Fig. 10). He expounded on this in three invited lectures[23] in Stockholm in February 1934 and later in his Hugo Müller lecture to the Chemical Society in London in March 1937.

The geochemical cycle treats the crust of the Earth, together with the associated atmosphere, hydrosphere, and biosphere, as a physicochemical system largely differentiated from the underlying mantle, although receiving from it energy in the form of heat and material in the form of magma. The cycle begins with the initial crystallization of magma, with the formation of igneous rocks, and with the addition of volatile matter to the atmosphere and hydrosphere (magmatic gases consist largely of H_2O, but with significant amounts of CO_2, H_2S, SO_2, and lesser amounts of other components such as HF and HCl).

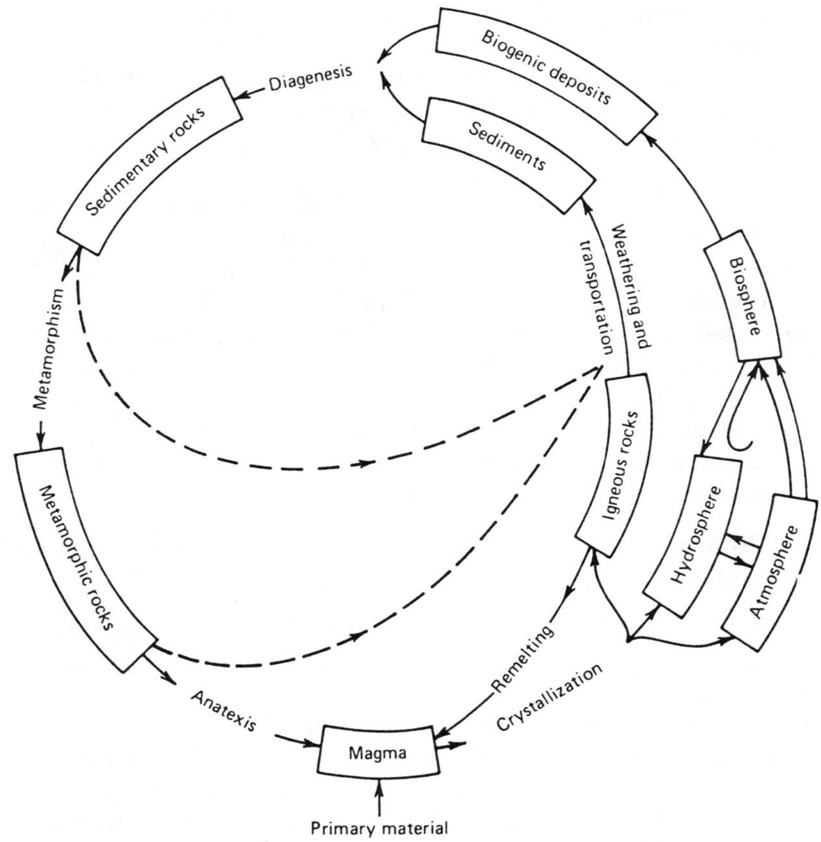

FIG. 10. The geochemical cycle in the outer part of the Earth.

The cycle proceeds through the weathering of igneous rocks and the transportation and deposition of the products as sediments, to form sedimentary rocks. During weathering some elements, such as the alkali metals and calcium, go into solution and are eventually delivered to the ocean. The ocean supports a vast amount of organic life, some of which results in the large-scale deposition of calcium carbonate, to form limestone and dolomite. Land plants convert CO_2 and H_2O into complex organic compounds, some eventually to form coal and oil. Sedimentary rocks, if buried deeply in the crust, react to heat and pressure to form metamorphic rocks; if sufficient heat is available melting may occur with the regeneration of magma. Goldschmidt's ultimate purpose was to quantify the geochemical cycle and to follow the course of the individual elements in proceeding through the different stages.

His work on crystal chemistry had already provided a fair understanding of the geochemistry of magmatic crystallization. He now proceeded to consider sedimentary processes. He pointed out that this part of the geochemical cycle was critically dependent on reactions taking place in water—weathering is essentially the reaction of water, together with oxygen and carbon dioxide, with rocks at the Earth's surface. He found that the behavior of an element in this environment could be predicted from its *ionic potential,* defined as Z/r, Z being the charge on the ion and r its radius, and illustrated this with a diagram (Fig. 11).

The ionic potential of an element largely determines its place of deposition during the formation of sedimentary rocks and is significant in all mineral–forming processes in an aqueous medium. It provides an explanation for the similar behavior of dissimilar elements as, for example, the tendency of the hydrated ions of bivalent beryllium, trivalent aluminum, and quadrivalent titanium to precipitate together during sedimentation. Elements of low ionic potential, such as sodium, calcium, and magnesium, remain in solution during the processes of weathering and transportation; elements with intermediate ionic potential are precipitated by hydrolysis, their ions being associated with hydroxyl groups from aqueous solution; and elements with still higher ionic potentials form anions containing oxygen which are usually again soluble. Hence the division of Fig. 11 into three parts: soluble cations, elements of hydrolysates, and elements of soluble complex anions. For elements of variable ionic charge, the ionic potential will vary: thus iron in the ferrous state is stable in solution, but oxidation to the ferric state greatly increases the ionic potential, and it is then precipitated as ferric hydroxide.

Goldschmidt likened the cycle of matter at the Earth's surface to the separation of the elements in a classical chemical analysis. The chemical differentiation that results is remarkable. The steps in this geochemical separation process are as follows:

1. Minerals that are especially resistant to chemical and mechanical breakdown collect as granular material. Of these the commonest

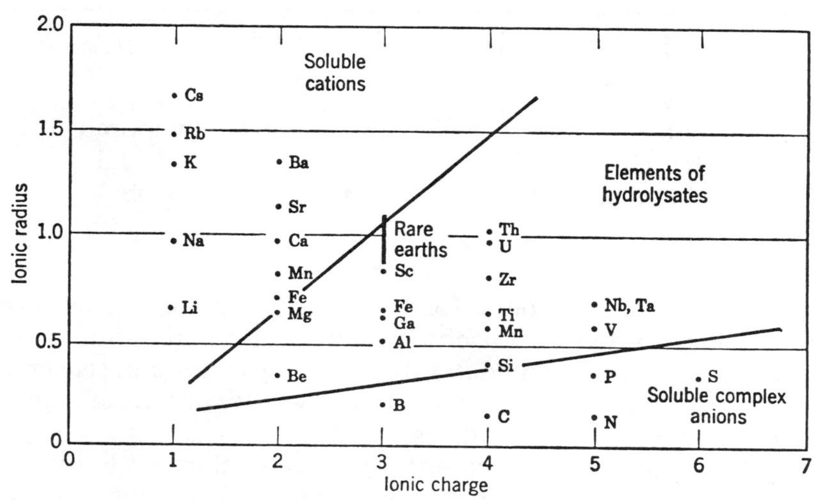

FIG 11. Geochemical separation of some important elements on the basis of their ionic potential (Goldschmidt, *Geol. Fören. Förh.*, vol. 56, p. 409, 1934).

is quartz, and the product is a quartz sand or sandstone showing a concentration in silicon with respect to the parent material. This may be compared to the separation of silica in the first stage of a rock analysis.

2. Accumulation of the products of the chemical breakdown of aluminosilicates, giving a mud consisting essentially of clay minerals. This results in the concentration of aluminum, and corresponds to the second step in a rock analysis, the separation of alumina and other easily hydrolyzed bases.
3. Along with the formation of clay sediments, but often separated in space and time, iron is precipitated as ferric hydroxide. In this process oxidation from the ferrous to the ferric state precedes precipitation by hydrolysis. Concentration of iron is the result, sometimes to the formation of iron ores.
4. Calcium is precipitated as calcium carbonate either by purely inorganic processes or by the action of organisms. Limestones are formed, and calcium thereby concentrated. This may result in practically quantitative separation of calcium, as in a chemical analysis. Limestone can be partly or wholly converted to dolomite by the action of magnesium–rich solutions, and magnesium thereby precipitated and concentrated together with calcium.
5. The bases that remain in solution collect in the ocean, from which they are removed in quantity only by evaporation, giving rise to salt deposits. The most important of these bases is sodium, but lesser amounts of potassium and magnesium also accumulate in sea water.

Goldschmidt depicted the chemical breakdown of a rock by weathering by the following scheme:[23]

Si	Al, Si, (K)	Fe	Ca, (Mg)	(Ca), Na, (K), (Mg)
↓	↓	↓	↓	↓
Resistates SiO_2	Hydrolysates, clay minerals	Oxidates $Fe(OH)_3$	Carbonates $CaCO_3$ $CaMg(CO_3)_2$	Evaporates $NaCl$ $CaSO_4$ $MgSO_4$ etc.

The scheme depicts the course followed by the major elements during sedimentation and gives a useful geochemical classification of sediments into *resistates, hydrolysates, oxidates, carbonates,* and *evaporates.* Goldschmidt recognized a further class, the *reduzates,* which include coal, oil, and sedimentary sulfides.

He then proceeded to quantify the sedimentary cycle by some ingenious calculations, i.e., to determine the total amount of igneous rock weathered during geological time and the amounts of the different sediments thereby produced. For each square centimeter of the

Earth's surface, there are 278 kg of sea water; and since sea water contains 1.07% sodium, the 278 kg contains 2.975 kg of sodium. The average sodium content of igneous rocks is 2.83%, and of sedimentary deposits approximately 1%. In the process of weathering a certain amount of the material is removed in solution, and Goldschmidt estimated that the mass of the sedimentary deposits is 0.97 of the igneous rocks that gave rise to them.[23]

Let X be the amount of igneous rock eroded per cm^2 of earth's surface.

Let Y be the amount of clastic sediments deposited per cm^2 of earth's surface.

Then: $Y = 0.97X$.

$$\text{Sodium content of igneous rock per cm}^2 = \frac{2.83}{100} \times X$$

$$\text{Sodium content of clastic rock per cm}^2 = \frac{1}{100} \times Y$$

$$\text{But the sodium content of ocean water per cm}^2 = 2.975 \text{ kg.}$$

$$\therefore \frac{2.83}{100}X - \frac{Y}{100} = 2.975$$

$$\therefore X = 160 \text{ kg/cm}^2$$

$$Y = 155 \text{ kg/cm}^2$$

He then used a similar type of calculation to determine the amount of calcium carbonate and dolomite in sedimentary rocks and obtained the following figures: 10.17 kg $CaCO_3$ and 4.37 kg $CaMg(CO_3)_2$. Hence the total amount of sedimentary rocks per square centimeter of the Earth's surface is 155 kg clastics, 10.2 kg limestone, and 4.4 kg dolomite.

Goldschmidt now proceeded to apply these results, and data on the abundance and distribution of individual elements, to a consideration of the geochemistry of the hydrosphere, specifically sea water, which comprises over 98% of the mass of the hydrosphere. He examined the balance between supply and removal of several elements in sea water. The basis of his comparison was the total amounts of different elements supplied by weathering and sedimentation during geological time. His calculations showed that about 160 kg of igneous rocks have been weathered for each square centimeter of the Earth's surface. Since there are 278 kg ocean water per square centimeter, approximately 600 g of rock have been weathered for each kilogram of water in the oceans. This 600 g of rock have therefore been the potential source of the dissolved matter supplied by weathering to 1 kg of sea water. Of course, only a part of this 600 g has actually dissolved and remained in solution. Goldschmidt drew up a "balance sheet" relating the potential supply of several elements in 600 g of igneous rock to the amounts present in 1 kg of sea water (Table 5).

Table 5. Balance of some elements per kilogram of sea water (Goldschmidt, *J. Chem. Soc. (London)*, p. 666, 1937).

Element	Supplied	Present	Percentage
B	2.4 mg	6 mg	250
Cl	0.29 g	19.3 g	6655
S	0.3 g	0.88 g	293
I	0.18 mg	0.05 mg	28
F	150 mg	1 mg	0.7
P	470 mg	0.06 mg	0.013
Na	16.8 g	10.7 g	64
K	15.0 g	0.37 g	2.5
Mg	12.6 g	1.3 g	10
Sr	250 mg	10 mg	4
Ca	21.6 g	0.42 g	2
Al	52.6 g	0.6 mg	0.0011
Si	160 g	1 mg	0.0006
Pb	10 mg	5 μg	0.5
Se	0.4 mg	4 μg	1
As	3 mg	20 μg	0.7

The most striking feature of Table 5 is that the concentrations of a few elements are far greater than can have been supplied by weathering. These are the common anions of sea water–chloride, sulfate, and borate. Either they were present in great amounts in the primeval ocean or they have been supplied largely from volcanic gases and thermal springs throughout geological time. Judging from the data on amounts and composition of igneous exhalations, the latter alternative is adequate to explain at least the major part of these "superabundant" elements in sea water.

Ionic potential is important as a measure of behavior of an element in aqueous solution, but the data in Table 5 show that other factors must be significant in the ocean. Thus, of the alkali elements, which all have low ionic potentials, only sodium remains to any extent in the sea; little potassium is present, because it is largely removed by reaction with clay minerals to form mica. The alkaline earth elements are also largely removed from sea water. Biological activity is, of course, responsible for the extraction of much of the calcium. The very small amounts of trivalent and quadrivalent elements in sea water evidently reflect their high ionic potentials, causing them to be almost quantitatively precipitated.

A remarkable feature is the very low percentage of certain poisonous metals and metalloids in sea water, including some, such as selenium and arsenic, which form soluble complex anions and therefore might be expected to accumulate. The quantities of these elements potentially supplied to the ocean during geological time are so great that a serious poisoning action would have resulted had not some process been active in eliminating them. Goldschmidt suggested that they have been removed from solution by a process similar to that used in practical medicine, that is, adsorption on freshly precipitated ferric hydroxide. He demonstrated the truth of this suggestion by finding that sedimentary iron ores show considerable concentrations of selenium, arsenic, and lead.

Table 6. Annual turnover of CO_2 per sq. cm. of the Earth's surface ($1\ \gamma = 10^{-6}$ g). Goldschmidt, *Geol. Fören. Förh.*, vol. 56, p. 415, 1934.

	γ
+ Addition of juvenile CO_2	3–6
+ Combustion of coal and oil	800
+ Respiration and putrefaction	ca. 40,000
− Photosynthesis	ca. 40,000
− Weathering processes	3–4
− New carbonaceous sediments	0.3–2

Goldschmidt chose to illustrate the geochemical cycle by drawing up a balance sheet for the yearly turn–over of carbon dioxide (Table 6 and Fig. 12). Carbon, although not one of the most abundant elements in the Earth, plays a most important role in geochemistry, because carbon compounds are essential for every known form of life. The geochemistry of carbon is closely linked with that of the other essential elements of organisms, especially hydrogen, oxygen, nitro-

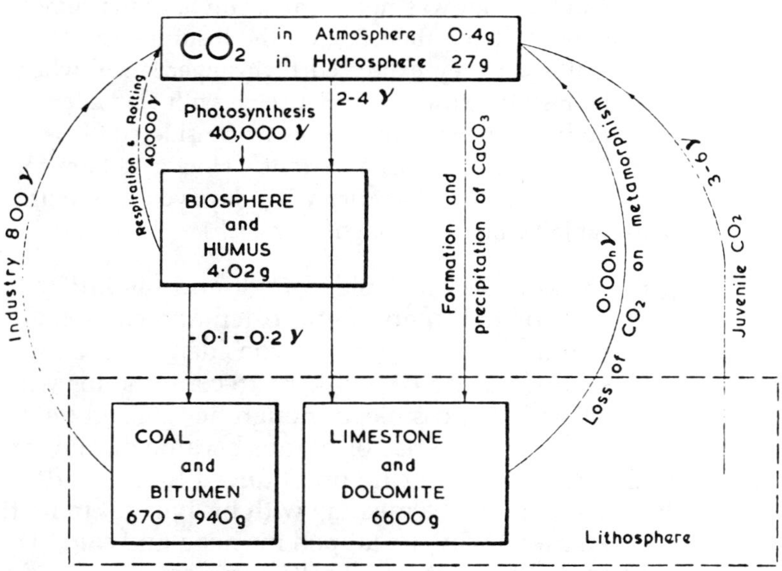

FIG. 12. The geochemical cycle of carbon ($1\gamma = 10^{-6}$ gram); after Goldschmidt, *Geol. Fören. Förh.*, vol. 56, p. 416, 1934).

gen, and sulfur. His balance sheet shows how overwhelmingly the cycle of carbon is determined by biochemical reactions.

He showed remarkable foresight when, in 1936, he anticipated the significance of man-made carbon dioxide emissions, in the following statement:[24]

> "The carbon cycle is of especial interest because it demonstrates the great significance that the industrial combustion of coal and other fuels has already had on the carbon dioxide content of the atmosphere. The amount of carbon dioxide which each year is added to the atmosphere by the combustion of fuels is two hundred times greater than that contributed by the world's volcanoes. This demonstrates that human activity in our time is a highly important geochemical factor" (translated from Norwegian).

What might he say today?

The peaceful and productive lives of the Goldschmidts, father and son, were fundamentally changed and their work interrupted by Hitler's accession to power in 1933. V. V. Shcherbina, who was working with Goldschmidt at that time, described the situation as follows:[25]

> "The Fascist putsch of March 1933 strongly influenced the work of the institute in a detrimental way. Several coworkers left the institute. I was a witness to the occurrences several days after the Fascist takeover, when a group of brownshirts with red bands and swastikas on their sleeves surrounded the institute and attempted to occupy it. "We don't fight with you" answered Goldschmidt's deputy Laves (Goldschmidt was not in the institute at that time). The Fascists wanted to hang their flag on the institute; this they did, and during this act they expressed their anger since, when they entered, the institute coworkers met them with cold silence. When Goldschmidt returned to the institute he considered the Fascist flag on his institute as a personal affront. "How glad I am for you, dear colleague" he said "that after a few days you will return to your comfortable Leningrad" (original English).

Following the Nazi takeover, Goldschmidt and his father, who had not previously belonged to any faith, ostentatiously joined the small Jewish community in Göttingen. He considered it his duty, in the interests of his students and co-workers, to continue his work at the university for as long as possible, although he realized he might be dismissed at any time. His sense of humor became grimmer than ever. When Hitler came to power he pronounced that "heads must roll". Goldschmidt appeared at a meeting with his hair cut more than usually short. When asked why, he tapped his head and said "Damit es besser rollen kann" ["So that it will roll more easily"].

Already by May, 1933, Goldschmidt's position was becoming difficult, as shown by the following letter from Adolf Windaus (1876–

1959, Professor of Chemistry in Göttingen, 1915–1944) to the Curator (Rector) of the University of Göttingen:[26]

> "I am grateful that you remain here, so that we have a man in Göttingen to whom one can turn for help and advice. Professor Goldschmidt has informed me in great distress that some members of the inorganic section of my institute have demanded a boycott of his lectures and laboratories. He named especially one Schorstein, a doctoral candidate of Professor Jander.
>
> I wish I could take serious action against this man. I believe, however, that this would only lead to my own departure, and I will try to remain here for as long as it is still possible.
>
> I ask your advice or your instruction as to how I should proceed. Till now I have always forbidden any sort of political activity within the institute; such a prohibition seems no longer possible" (translated from German).

Goldschmidt was to have been a guest at the International Geological Congress in Washington, D.C., in July 1933, and to have given two invited lectures, but the political developments in Germany prevented this. Later he was invited to be the George Fisher Baker Non–Resident Lecturer in Chemistry at Cornell University but was unable to take it up. He never visited North America, although he had many friends and acquaintances in the United States and Canada, and extensive business dealings with the Harbison–Walker Refractories Co. in Pittsburgh.

By 1935 his situation was becoming unbearable, as indicated by a letter from the NSDAP (Nazi) Gauleitung (local administration) to the University of Göttingen:[27]

> "The Gauleitung must deny the request of Professor Dr. Goldschmidt for permission to give a lecture at Lund University in Sweden. Today it is obvious that when a distinguished scientist gives a lecture in a foreign country he will be questioned concerning political developments in Germany. On account of his Jewish mentality Professor Goldschmidt is unable to consider these matters as we would wish, and his judgement is thereby distorted. For that reason the Gauleitung cannot approve Professor Goldschmidt's proposed visit to Sweden" (translated from German).

The conditions under which Goldschmidt left Göttingen in 1935 were explained in his letter in response to a query by a friend, Dr. J. Shearer of the Physics Department of the University of Western Australia. Goldschmidt wrote:

> . . . "It is not true that life was rendered unendurable and work impossible for me at Göttingen. The authorities of the University of Göttingen during all my stay in Germany always treated me in a most pleasant and agreeable manner, giving me any facilities of scientific work, nor had I any difficulties with assistants, students, colleagues, officials or others. But you will understand that the gen-

eral attitude against non–Aryans was felt by me as a very great misfortune, even if not in any case used against me personally. In spite of my feelings, which you can understand, I have felt it to be my duty towards science, towards my coworkers at the institute and towards the university to keep my work at Göttingen going on, through the years 1933, 1934, and until summer 1935, an attitude which was encouraged in every manner by all authorities which had to do with my work. I can mention that funds for my laboratories even in that time generously were placed to my disposal, and teaching and cooperation with my colleagues went on in a most friendly manner. However already in 1933 I found it to be my duty to resign from the chairmanship of "Gesellschaft Deutscher Naturforscher und Aertze, Naturwissenschaftliche Hauptgruppe" (corresponding to your "British Association") to make it clear for all concerned that I did not wish to consider myself a representative of science at the same time when my colleagues were driven from their positions, which like myself were non–Aryans.

Then in spring 1935, on May 1 there was placed in Göttingen a big sign board, not far from the road to my institute, with the inscription "Jews not desired". I gave notice that I was resigning my professorship if such board was not taken off, as I could not reconcile my presence in Göttingen with such an open attack against Jews in the same town. Before 24 hours had gone the thing had been taken off. However several months later (in August) a signboard of the same kind was reerected, that time just opposite my institute. I renewed my action of protest, and that time it was not removed even at my request, as such boards at the same time had been placed in all German towns by direct order of the uppermost party rulers. Consequently next day, August 11, as an ultimate protest I resigned my position, and designated my intention to leave the country.

I may mention, that all negotiations have been made in a most friendly manner and that the authorities of the University in every conceivable manner supported me in order to make agreeable terms for my emigration back to Norway when they learned that my decision could not be moved. However, as you will understand, under existing financial restrictions in Germany, my father and I had to leave at that time all our financial resources in Germany, but we were allowed to take back to Norway all our furniture, library and my instruments for research. Later on I even have obtained uncommonly favorable conditions—as compared with other cases—for transferring even a certain percentage of my financial resources back to Norway. Of course my emigration in 1935 did involve a great financial risk for me, resigning my position and having no other position in another place in expectance.

That my departure from Göttingen, even in being an action of protest from my side, was effectuated in a friendly spirit you may

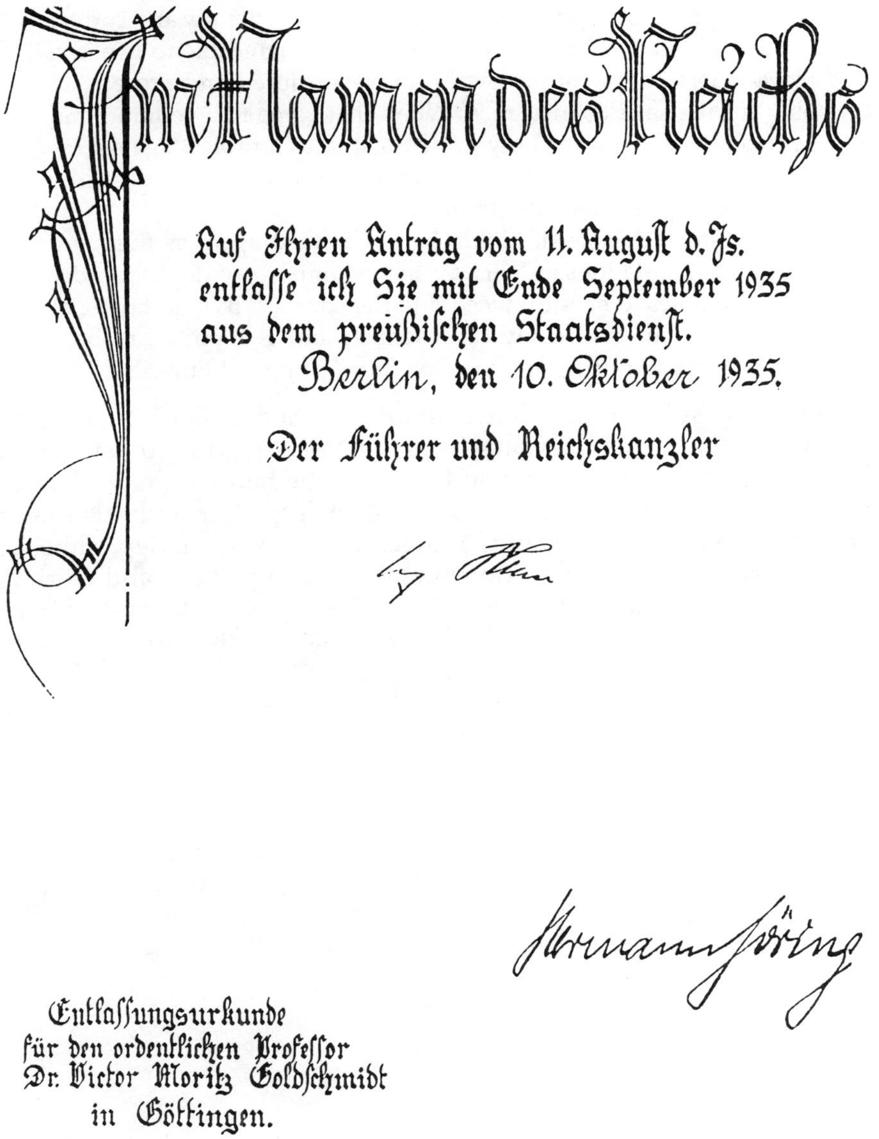

Im Namen des Reichs

Auf Ihren Antrag vom 11. August d. Js. entlasse ich Sie mit Ende September 1935 aus dem preußischen Staatsdienst.

Berlin, den 10. Oktober 1935.

Der Führer und Reichskanzler

Entlassungsurkunde
für den ordentlichen Professor
Dr. Victor Moritz Goldschmidt
in Göttingen.

FIG. 13. Dismissal letter for Professor Dr. Victor Moritz Goldschmidt, signed by Hitler and Göring.

learn from the fact that my whole laboratory, including students, assistants, artisans, officials gave a festival for me before I left, and that ever since my departure there has been friendly cooperation between my former institute at Göttingen and my new institute in Oslo, even exchange of assistance and scientific cooperation.

Of course I feel very heavily the injury done to my colleagues, which are non–Aryans as am I, but I feel it necessary to say the

> fact that I myself always have been treated in a friendly and correct manner by all authorities, even if it was not agreeable to live in a place where people of my own race were injured and prosecuted only for the sake of their race. When that came to a certain limit, as told above, I resigned my professorship as a protest against iniquity. I feel sure that anybody in my case would have done the same, perhaps even earlier than I did.
>
> My small dog, Bazi, had to be left at Göttingen, as there are very close regulations in Norway against imports of dogs, in order to avoid rabies to be introduced into the country. Bazi is staying at Göttingen with the family of a dentist, and is reported to be a happy, but still rather naughty dog there" (original English).

On August 25, 1935, Goldschmidt travelled to Berlin, where he received a Norwegian passport from the Norwegian Embassy (when he accepted his professorship at Göttingen he had automatically become a German citizen, as a civil servant). Together with his father and Miss Brendingen, he left Göttingen for Norway on September 6. He later received an official document, signed by Hitler and Göring. . . . "At your request of August 11 I release you at the end of September 1935 from service to the Prussian state" (Fig. 13).

8 / Oslo: 1935–1942

WHEN GOLDSCHMIDT TOGETHER with his father and Miss Brendingen, returned to Oslo on September 9, 1935, he was practically penniless. He was permitted to take his household furniture, his scientific library, and some of his instruments from Germany, but was limited to 60 Reichsmarks in cash. His bank accounts were frozen, and his house and other real property sequestered. Later, part of his money was transferred to Norway as a result of an outspoken letter from his successor at Göttingen, Dr. F. K. Drescher–Kaden, to the President of the Reichsbank, Dr. Hjalmar Schacht; his house and all his other property were confiscated without compensation. Fortunately, his work on the development of olivine refractories was bearing fruit; from 1935 the Harbison–Walker Co. of Pittsburgh paid him $2000 a year for the rights to his United States and Canadian patents.

His longtime assistant at the Raw Materials Laboratory, Kristoffer Stenvik, had arranged temporary accomodation for Goldschmidt and his father; Miss Brendingen went to live with her brother. It was not until early 1936 that he, his father, and Miss Brendingen were able to re–establish their household in the upper floor of a large house[1] in the suburb of Holmenkollen, with a magnificent view of the city and fjord (Plate 31).

He received many letters from friends and colleagues welcoming his return to Norway, including one from his old student, Tom. Barth, who was working at that time in the Geophysical Laboratory of the Carnegie Institution of Washington. In a letter dated October 30, 1935, Barth wrote:

> "From Zachariasen I have learned that you have now returned to Oslo, and he relayed your friendly greetings to me, which I greatly appreciate. Some days ago I also received a letter from Saeland[2] informing me that it is finally arranged with the Commerce Department that I shall receive 300 kr. per month in addition to my docent salary. I know of course that already a year ago you recommended me warmly to the Commerce Department, and advised

me to establish a connection with raw materials investigations in Norway; now that this has been arranged, I am aware that it is largely due to your recommendation, and believe me when I say how much I appreciate your help. I know that it is my wife's and my parents' greatest desire that I return to Norway, even though I know that conditions will be very different to those I have enjoyed in this country; however, thanks to the Commerce Department's stipend we will be reasonably placed economically, and I have good possibilities to obtain research funds. In addition, I understand that the docent position is probably transitional, and I can hope for promotion to professor in the near future.

From Saeland's letter I understand that you are now engaged in raw materials work, and that possibly I will have the opportunity of cooperating with you. This would please me greatly—it would be a great experience to work with you again" (translated from Norwegian).

Unfortunately, as will be seen, these hopes for friendly cooperation were not to be realized.

During the Göttingen years Goldschmidt had continued as an advisor to the Raw Materials Laboratory, and on his return the Commerce Department reinstated him as Director and he moved into his old quarters at the Geological Museum—much to the astonishment of the Rector of the University, who learned it first from the newspapers. Fortuitously, Dr. Jakob Schetelig,[3] who in 1917 had succeeded W. C. Brøgger as Professor of Mineralogy and Petrology and director of the Geological Museum, died on October 17, 1935, after a long illness. The position was declared vacant, Goldschmidt applied for it, and was duly appointed in March 1936.

The previous years had been marked, in Norway as in the rest of the world, by economic depression. The university suffered, student numbers in geology dwindled almost to zero, and the upkeep of the Geological Museum had been seriously neglected. Goldschmidt attacked these problems with gusto. He obtained government funds for the rehabilitation of the museum and his laboratories. He installed the scientific equipment which he had been able to bring from Göttingen, and purchased more, some with his own funds. In particular, he arranged for his Göttingen colleague, Dr. Mannkopff, to construct new spectrographs for him, and to come to Oslo in November and December 1936 and install them. Student numbers increased in 1936 and in the following years, and research activities in the museum took a new lease on life. Besides his university work, Goldschmidt devoted much time and effort to the development of olivine refractories, for which he held patents both in European countries and North America, and from which he received considerable royalties.

The Barth family returned to Oslo from the United States in February 1936, and Goldschmidt arranged for them to rent the ground floor of the Holmenkollen house. However, it was not long before friction arose between the two men. Barth has written the following account of his initial conflict with Goldschmidt:[4]

> "One day at the end of October 1936, Professor Goldschmidt rang my doorbell at 7.50 a.m. and immediately asked why, without consulting him, I had applied for money for the investigation of the utilization of different rock types for road metal. I answered that I had never thought he was interested in research on road metal, that the application was urgent and left no time for such a conference, and it was a problem that had interested me for more than 15 years, so it was nothing new on my part.
>
> Goldschmidt, however, would not accept my explanation, and accused me of going behind his back. For example, I had—also without asking him—during the past summer made geological studies in Jotunheimen, which was his old area, and he could not accept that I stole it from him. In addition, I had associated with his enemies, Saeland and Holtedahl.[5] Holtedahl had stolen the Alunsjø area from him while he was in Göttingen.
>
> He further forbade me to cooperate or confer with Dr. Arne Bugge[6] or Dr. C. Bugge[7] in my office at Tøyen. If he saw any of them come he would order the watchman not to admit them.
>
> Finally he said that I would have to choose between him and Holtedahl and Saeland.
>
> There was thus open enmity between Goldschmidt and me for a time. To put an end to this, a week or two later I went up to him at Tøyen and said that I felt it was unfortunate that he had taken umbrage in this way, but added that he must understand that he could not control with whom I should associate. He replied that he was not interested. I then asked if he would retract what he had said about Saeland and Holtedahl. This he would not do, and said that I would soon learn the true nature of these gentlemen. I answered that he must understand that I could still be loyal to him even if I associated with Saeland and Holtedahl. Nevertheless, he insisted that I would have to choose between him and Holtedahl and Saeland" (translated from Norwegian).

Barth felt short-changed in the amount and quality of the quarters he was allotted in the museum, whereas Goldschmidt claimed that he was more than generous in meeting Barth's requests. At this time the university was developing a new campus at Blindern, on what was then a farm on the northern outskirts of Oslo. The first building, for physics and chemistry was completed in 1934. Some space in this building was made available for the geology courses. In 1937 Barth was appointed Professor of Crystallography and Petrology, and the Mineralogical Institute, which had been in limbo after Goldschmidt went to Germany in 1929, was reactivated with Barth as Director and was established on the new campus.

At Christmas 1936 the Goldschmidt household was completed by the acquisition of a cocker spaniel, "Highland Bob," to replace the dachshund Bazi which had been left behind in Göttingen. Goldschmidt presented Highland Bob to his father, as a companion for his daily walks. Highland Bob was a pedigree dog, winning prizes at Oslo Kennel Club shows and sought out for stud purposes (fee 75 kroner) (Plates 33, 36).

In January 1937 Goldschmidt was hospitalized for two weeks with nephritis. He made a good recovery, but the kidney trouble was to recur and plague him in his later years. In March, he spent a week in London at the invitation of the Chemical Society, to deliver the seventh Hugo Müller Lecture, *The Principles of Distribution of Chemical Elements in Minerals and Rocks.* The published paper[8] is a lucid summary of the geochemical work of Goldschmidt and his co-workers during the previous fifteen years, and the best available exposition in English prior to the publication of his book *Geochemistry* in 1954. At this time he completed the manuscript of the ninth part of the *Geochemische Verteilungsgesetze der Elemente,* entitled *Die Mengenverhaltnisse der Elemente und der Atom-Arten.* It was submitted for publication in May and was published in March 1938.

Goldschmidt referred to it as his "Ninth Symphony." In it he critically reviewed all the available data on the abundances of the chemical elements, mostly from the Göttingen institute but including some from other sources. He provided three summary tables: abundances of the chemical elements in igneous rocks, in meteorites, and in the solar atmosphere. From a combination of meteoritic and solar data, he arrived at a table of cosmic abundances (Fig. 14). The significant features may be summed up as follows:

1. The abundances show a rapid exponential decrease for elements of lower atomic number (to about atomic number 30), followed by an almost constant value for the heavier elements.
2. Elements of even atomic number are more abundant than those of odd atomic number on either side (the Oddo–Harkins rule). It had been clearly recognized in the case of the rare earth elements by Goldschmidt and Thomassen in 1924.
3. Only 10 elements–H, He, C, N, O, Ne, Mg, Si, S, and Fe—all with atomic numbers less than 27—show appreciable abundance; of these, hydrogen and helium far outweigh the other eight.
4. There is a pronounced abundance peak at atomic number 26 (Fe) and smaller peaks at several other heavier atomic numbers.
5. Elements 3, 4, and 5—Li, Be, B—show remarkably low abundances; this had been noted by Goldschmidt already in 1926[9] and ascribed to a probable instability of their nucleii, which was later confirmed.

Goldschmidt realized that elemental abundances were linked to properties of the nuclear structure. The nucleus of an atom can be

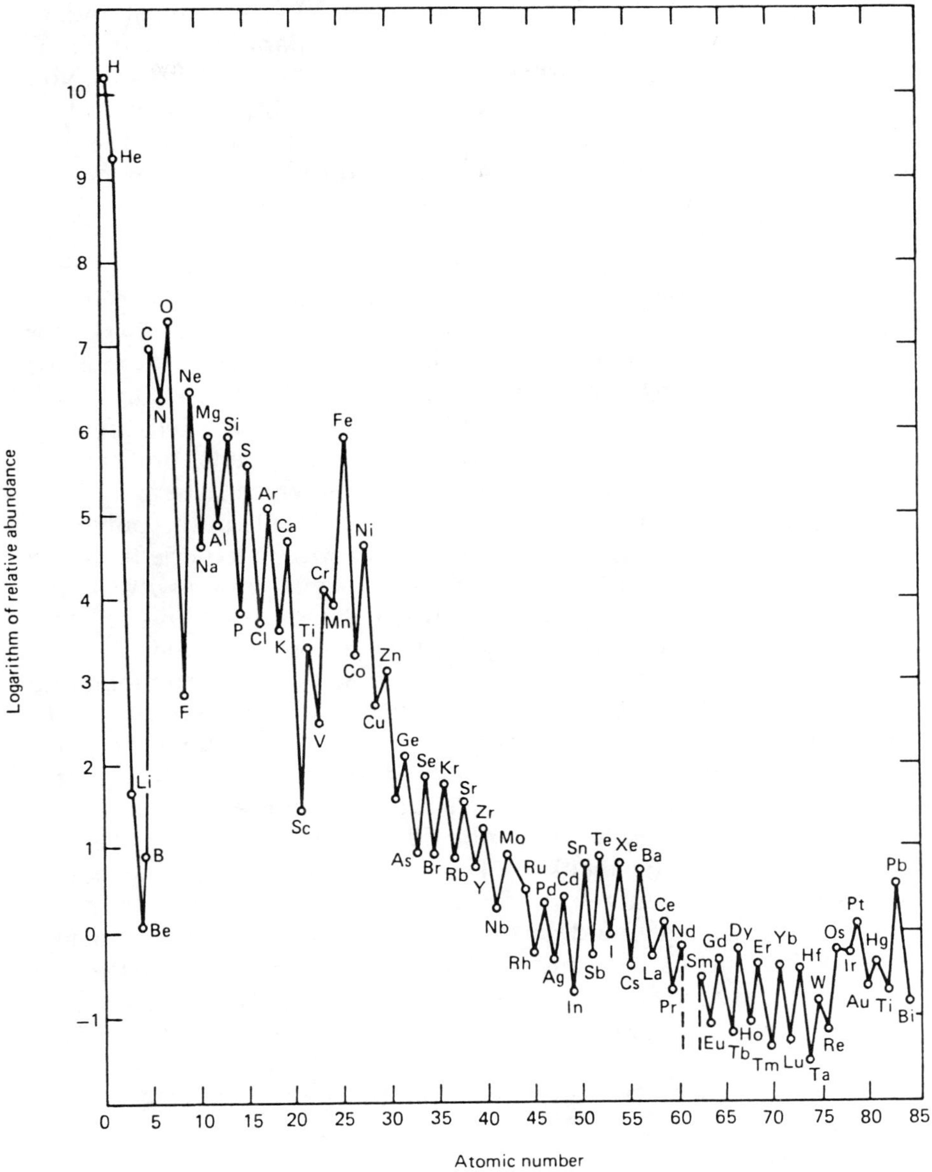

FIG. 14. Relative abundances of the elements, normalized to Si = 10^6 atoms, plotted against atomic number (after Goldschmidt, *Geochemische Verteilunggesetze der Elemente IX*, p. 117, 1938).

considered as made up of positively charged protons (Z) and uncharged neutrons (N), each of unit mass. An element is characterized by its proton number, equal to the atomic number. Individual elements have a fixed number of protons in the nucleus, but may have a variable number of neutrons–specific nuclei with different numbers of neutrons are known as isotopes. Goldschmidt noted that certain neutron numbers had unusual abundances; these were 30 (the most abundant isotopes of Mn, Fe, and Ni); 50 (the most abundant isotopes of Sr, Y, and Zr); 82 (the most abundant isotopes of Ba, La, Ce, Pr, and Nd); and 108 (the most abundant isotopes of Hf and Ta). He realized that these must be nuclei of unusual stability, and this reflected the processes by which complex nuclei have been formed from primary particles, presumably in the interiors of stars and supernovas.

Goldschmidt's cosmic abundances table has been the basis of subsequent theories of atomic structure and the origin of the elements. It provided the foundation for two subsequent Nobel prizes in physics: (1) to Eugene Wigner (who coined the term "magic numbers" for these unusually abundant nuclei), Maria Goeppert Meyer, and Hans Jensen in 1963 for their theory of nuclear shell structure, in which the magic numbers are those for which specific nuclear shells are fully occupied; and (2) to William Fowler in 1983 for the theoretical and experimental studies of the nuclear reactions which determined the formation and abundance of the chemical elements in the universe.

In his "Ninth Symphony" Goldschmidt speculated briefly on the possible existence of trans–uranium elements, i.e., those with atomic numbers greater than 92. He had actually demonstrated his remarkable prescience in this regard some years earlier, in an invited paper to a Russian conference held on February 8, 1934, to commemorate the centenary of the birth of D. I. Mendeleev; the contributions to this meeting were not published until 1937.[10] In his paper he speculated on the chemical properties of the trans–uranium elements, and he argued from theories of atomic structure that they would belong to an analogous series to the lanthanides, which he termed the thorides, after thorium (element 90). They are now known as the *actinides,* after actinium (element 89), the first member of the series. He predicted that these elements would have a principal valency of four, and secondary valencies of three and five. He theorized that their ions would show a contraction with increasing atomic number similar to the lanthanide contraction. From his previous determinations of the ionic radii of thorium and uranium (Th^{4+} 1.10Å, U^{4+} 1.05Å), he estimated the radii of the quadrivalent ions of the trans–uranium elements, as follows:

AN	93	94	95	96	97	98
r	1.03Å	1.01Å	0.99Å	0.98Å	0.96Å	0.94Å

In a remarkable confirmation of his predictions, the values subsequently determined for these elements are:

AN	93	94	95	96	97	98
r	1.01	1.00	0.99	0.99	0.97	0.96Å

After the identification of neptunium (93) and plutonium (94) at the University of California in Berkeley in 1940, Goldschmidt returned to this subject and in January 1942 he published a short paper in

Norwegian pointing out how his prophesies regarding the chemistry of these elements had been confirmed.[11]

Goldschmidt was an omnivorous reader, or perhaps I should say scanner of the chemical and geological literature. I recall seeing him peruse the latest issue of *Chemical Abstracts,* running his finger rapidly down the columns, pausing only occasionally to mark or note an abstract for further study. And he had a unique capacity to retain and recall this information. An interesting example is his correspondence with H. F. Harwood[12] and Arthur Holmes[13] regarding their publication *The Volcanic Area of Bufumbira,* a 300-page monograph published in 1937 by the Geological Survey of Uganda (Memoir 3). Writing to Harwood on September 29, 1937, Goldschmidt commented:

> "I am sorry to inform you that I consider some of the spectrographic data on page 52 to be considerably wrong. Especially I wish to draw your attention to the fact that your spectrographist reports scandium to be absent, that is said to be less than 0.0001%. From my own work on the geochemistry of scandium I can confidentially tell you that such rocks as the six rocks given in the table on page 52 *must* contain considerable amounts of scandium. The amount of scandium must certainly be at least about 5 times larger than the upper limit given in your table, probably in the pyroxenites even 50–100 times larger than your upper limit. Now, I am rather sorry that the laws of geochemistry seem to be violated by the results in your book. May I make the following proposition to you, which I think to be a fair one? I propose a wager on this matter: I am betting two guineas for each of the six rocks, that they contain appreciable amounts of scandium, at least 0.001% of Sc_2O_3. For each of these rocks, which contains not so much Sc_2O_3, I am apt to pay two guineas, if I lose. For each of the rocks, which contains as much as 0.001% Sc_2O_3 or more, I propose that you or your co–workers pay the same amount per rock. The total amount collected jointly may be given to a very good cause, that is the Society for the Protection of Science and Learning (formerly Academic Assistance Council), 6 Gordon Square, London WC 1" (original English).

Harwood referred this letter to his co–author, and Holmes replied on October 11, 1937:

> "I have received from Harwood a copy of your letter to him, with the suggestion that I should write to you on the point at issue. Your challenge is a beautiful illustration of the common belief that all the English are very wealthy! It also implies that we doubt the validity of what you say. Like Harwood, I am quite prepared to accept your judgement, particularly as you point out that the information on the subject was published only in 1936. Dr. Hitchen did the work for me in 1934, and could not be expected to obtain

results that required a specialized technique not then known to him. Dr. Hitchen is now in Kenya Colony and can therefore take only an academic interest in the problem for the present. Since the problem is evidently of unusual interest to you, I expect you would be glad to settle the matter for your own satisfaction by direct tests. If so, please let me know and I shall be happy to send you small samples of the rocks tested by Hitchen."

Goldschmidt replied on October 19, 1937:

"May I at first be allowed to settle a slight misunderstanding? I had never the opinion that all the Englishmen are wealthy, but I had the opinion that the excellent scientists of your country have usually a very firm belief in the correctness of the numerical data given in their own publications. As to the technique of Mr. Hitchen I wish also to set right your belief. The most convenient technique of making scandium determinations by means of spectrography has been published by me in 1931 in *Nachrichten der Gesellschaft der Wissenschaften zu Goettingen*, p. 257–279, a publication available for instance at the excellent library of the Geological Society, London. . . . Perhaps I may mention that not only the data on scandium in your memoir are entirely wrong, but several of the other spectrographic data of the same table seem to be somewhat unbelievable. Of course, one might consider that it does not matter much if incorrect data on geochemical topics are introduced in geological descriptions of remote territories, but what I fear is the danger that an accumulation of unbelievable spectrographical data may bring the excellent methods of analytical spectrography into disrepute. Therefore, I consider it necessary to have these data corrected, and I am glad to accept your offer to let me get some small samples of the six rocks involved."

Holmes sent these samples to Goldschmidt, who reported as follows:[14]

"These (samples) were spectrographically analysed in my laboratory in Oslo by L. W. Strock and A. Kvalheim and have the normal amount of scandium for such rocks, up to 80g/ton Sc_2O_3 in the pyroxenites and up to 5g/ton in the kimberlite."

An example of Goldschmidt's capacity to take offence where none was intended is the case of Harry Berman. Berman was assistant to Professor Charles Palache (1869–1954), Director of the Harvard Mineralogical Museum. He was a visiting researcher at Göttingen from September 1932 to April 1933, working with Goldschmidt and other colleagues on the geochemistry of the alkali metals. On his departure Goldschmidt wrote to Palache, whom he knew from a visit to Norway in the summer of 1924: "(Berman) is a clever man, especially he is a very good connoisseur of minerals and mineral paragenesis." This favorable evaluation changed abruptly after Berman published "The Constitution and Classification of the Natural Silicates" (*Am. Mineral.*, vol. 22, p. 342–408, 1937). Goldschmidt wrote to Professor Linus Pauling as follows (July 6, 1937):[15]

> "Some time ago I have read in the May issue of the *American Mineralogist* a so-called classification of the silicates by a Mr. Berman of Harvard University. As I understand from some American colleagues that the article is probably intended to be the basis of the chapter on silicates in the forthcoming new edition of Dana's hitherto very well renowned handbook of mineralogy, I dare to direct your attention to the matter.
>
> I really very seldom have seen a general article on structures being based on so much half understanding and with so many haphazard conjectures, and with so much ignorance concerning the really important work on silicates.
>
> Might it not eventually be worth while to direct the attention of the responsible editors of Dana's handbook to this matter? It is a matter of importance that so important a matter as the structure and classification of silicates must be presented by some competent person.
>
> Another most astonishing feature of that article is that the author begins his story by telling that no one ever has made an attempt to classify all silicates on the base of structural theory. He seems not to be aware of the monograph of E. Schiebold, published 1932–33 in *Ergebnisse der exacten Naturwissenschaften*, vol. XI and XII. Even though Schiebold's work is not a masterpiece, it is manyfold better than Berman's.
>
> I have met Berman 1932–33 at Goettingen, and I think he is not quite the personage to be entrusted with such literary work which demands very thorough scientific knowledge and intelligent judgment" (original English).

Goldschmidt wrote to Pauling rather than directly to Palache, presumably because Pauling was the leading American authority on silicate structures. Pauling forwarded the letter to Palache with the comment

> "The article by Berman seems to me to be unsatisfactory in several respects although not quite so bad as Professor Goldschmidt suggests. . . . The article does not give a satisfactory idea regarding the historical development of our modern understanding of silicate structure. For example, in the section beginning on page 345 on "Results of the X-ray Analyses of Silicate Structures," the reference given is to Bragg, 1930, with no mention of the investigators who formulated these principles."

This I think is the key to Goldschmidt's displeasure. He was extremely sensitive to any lack of appreciation of his own work or that of his colleagues. Since both he and Pauling had been in the forefront of crystal structure work in the 1920's, it was natural for him to turn to Pauling for support in this matter.

Palache replied to both Goldschmidt and Pauling in letters dated August 25, 1937. To Goldschmidt he wrote:[15]

> "I cannot but wonder whether you fully realize the main purpose of this paper. As a mineralogist you surely believe in the attempt to classify minerals, and your own work has contributed so conspicuously to the understanding of mineral groups that I know you must be in sympathy with any attempt to improve classification. I am sending you a copy of my reply to Dr. Pauling's letter in which I have attempted to give the point of view of the editors of the new edition of Dana. Not having heard from you directly, I do not know how much of this will meet your objections to Dr. Berman's work. I cannot believe, however, that he is 'ignorant concerning the really important works on silicates.' This work was the fruit of extremely careful study of all existing modern silicate analyses and its results were discussed very fully with Dr. Warren whose work in the field of silicate structure is, as you know, authoritative. It had Dr. Warren's full approval."

As a last word on this matter I may quote from C. S. Hurlbut's memorial to Harry Berman[16]:

> "His doctoral thesis on the constitution and classification of the natural silicates was based on the meager structural data in the early thirties. A large percentage of the silicates studied since that time have been found to fall in the exact place in the classification in which he placed them, thus showing his vision and proving that the system was well grounded."

Lester Strock, who had returned to the United States from Göttingen in November 1935, rejoined Goldschmidt in Oslo in May 1937, accompanied by his wife and year–old daughter, and stayed until December 1938. Goldschmidt entrusted him with the task of setting up an optical spectroscopy analytical facility similar to that in Göttingen, and in training assistants. One of these was Aslak Kvalheim,[17] appointed as chemist in the Raw Materials Laboratory in October 1937; he became a close friend and associate of Goldschmidt and directed the laboratory after Goldschmidt's escape to Sweden in December 1942.

Throughout his life Strock remained a devoted disciple of Goldschmidt. Whenever we met we would indulge in nostalgic recollections of our days in his institute. In 1972, The Geochemical Society instituted the Goldschmidt Medal as its premier award and Strock endowed it with a monetary stipend.

Goldschmidt's father Heinrich died suddenly of a heart attack on September 20, 1937, a few weeks before his eightieth birthday. It was a personal tragedy for Goldschmidt, since he had always been close to his father, and they had lived together throughout his life. Paul Rosbaud wrote:[18]

> "I went to stay with him shortly after his father's death. He had aged, his health had deteriorated, and he was embittered and tired. But he went on with his scientific and his relief work with almost inexhaustible energy. He spent more than he could afford in helping other people and managed to bring an aged aunt and a niece–still a child–from Germany to Oslo, and the niece from there to the U.S.A. He still enjoyed his food and an occasional drink and he maintained his old-fashioned courtesy. His humor had become more whimsical, more grim, and sometimes even caustic. I remember one day when I asked him to come with me to take some flowers to the crematorium as a tribute to his parents. We stood reverently in front of the urns, all made of beautiful green Norwegian olivine, two with the ashes of his parents and a third, empty and destined to hold his own, and finally V. M. remarked dryly: "Ja, Ja, the whole family in magnesium orthosilicate."[19]

The aged aunt was Mrs. Greta Koehne, the widow of the brother of Goldschmidt's mother. In 1938 she was living in Berlin, and he was able to obtain a Norwegian residence permit for her. She arrived in Oslo in January 1939 and lived with him until her death in November 1940. The niece was actually a grandniece, Erika Elisabeth Sarah Schulhof, born in Vienna on December 28, 1928; she was the daughter of Dr. Friedrich Anton Schulhof, the son of a sister of Goldschmidt's father. Through Goldschmidt's efforts she was able to leave Vienna for Norway in May 1939, and from there to Scotland, where she was educated at a convent in Aberdeen. Goldschmidt saw her often during his stay in Aberdeen in 1943–44, and they corresponded frequently until his death. After World War II she emigrated to the United States, where she had an aunt, Mrs. Marie Treuer, in Yellow Springs, Ohio. Her parents were deported to Poland in the autumn of 1941 and presumably perished.

Goldschmidt was also instrumental in obtaining a Norwegian entry permit for a cousin of his mother, the widow Stephanie Hirsch (born October 10, 1875) from Berlin. She came to Norway in January 1940 to await a visa for the United States, where her son, Dr. Felix Hirsch, was librarian at Bard College at Annandale-on-Hudson, New York. Because of the war Mrs. Hirsch was unable to proceed to the United States and she lived at Lillehammer, a country town about 100 miles north of Oslo, on money sent her by her son. After the United States entered the war the American payments were suspended, and Goldschmidt supported her financially. After the Jewish persecution began in 1942 he wrote to her several times, telling her to contact him if problems arose. On November 28, 1940 he received a postcard from her, saying that she was being sent from Lillehammer, saying goodbye and thanking him for his help. The postcard was presumably sent on Thursday morning, November 27, 1940 shortly before her departure. She was sent to Oslo by train on that morning, arrived in Oslo about 2 p.m., and with other Jewish prisoners taken directly to the harbor and shipped to Poland on the prison ship "Donau," which left at 3 p.m. She was never heard from again.

On January 27, 1938, Goldschmidt had his fiftieth birthday, entering what was to be the last decade of his life. He made some typical comments in a letter to a friend in Göttingen:

> "You will be interested to know that my fiftieth birthday was celebrated with much greater fuss than I had expected. I was visited at my home by distinguished delegations—from the university, led by the dean, from the Geological Survey, the presidents of the Geological Society and the Mining Engineers Association, a high official from the Ministry of Commerce, together with museum staff and students. And the newspapers and technical journals published the finest obituaries. I felt like the Emperor Charles V, who witnessed his own burial service" (translated from German).

In the same letter he comments on his working habits:

> "I work from 7.30 in the morning often to 2 a.m. the next day, with only a few minutes for meals, weekdays and Sundays, on my university responsibilities, on my industrial interests, and in the preparation of manuscripts."

And on a visitor:

> "The day after tomorrow I expect an out-of-town guest. Our university will have a large honorary graduation ceremony, and one of the honorary doctors will stay with me, a colleague from Helsingfors who in his youth did some of his most renowned work with me." (This was Professor Pentti Eskola.)

In 1938 Goldschmidt proposed his old student, Gulbrand Lunde, who was then Director of the Canning Industry Laboratory in Stavanger, for membership in the Videnskaps-Akademi (the Norwegian equivalent of the Royal Society). This was opposed by Barth, who cast doubt on the originality of some of Lunde's scientific work. Lunde successfully refuted these objections, and was duly elected, but the incident added to the growing hostility between Goldschmidt and Barth. The summer of 1939 saw a complete break between the two men. Barth had arranged to spend some months at the Geophysical Laboratory in Washington, D.C., starting on September 1. Barth proposed to sublet his half of the Holmenkollen house to a German diplomat, Dr. Ulrich Noack. Noack was a historian engaged in writing a book on Norway and Nordic history, and was attached to the cultural section of the German legation in Oslo. He was a member of NSDAP, the German Nazi party. Goldschmidt was furious that Barth would invite a German official to occupy part of the house. The situation was resolved by Goldschmidt agreeing to pay Barth's share of the rent during his absence.

In 1938, Goldschmidt acquired a small but remarkable piece of real estate, the islet of Låven in Langesundsfjord. Langesundsfjord is a large inlet on the west side of the Oslofjord, famous to mineralogists the world over for its nepheline syenite pegmatites and their wealth of rare minerals, some of them unique to this region. In 1890, Brøgger published a classic 898-page monograph (it was the complete vol. 16 of the *Zeitschrift für Kristallographie*) on the region, with descriptions of 73 minerals, many of them new to science. Goldschmidt visited the region in 1910 in the company of Brøgger and Paul von Groth, the foremost crystallographer of that time, and returned again and again with parties of students and foreign scientists. Låven is only 60 meters long and less than 10 meters high, barren of vegetation, but contains many of the minerals for which the region is famous. In 1937, the Christiania Minekompani, which owned the island but never exploited it, offered it to the university for 1500 kroner (about $500), but the university was unable or unwilling to buy it. Goldschmidt bought it to preserve it and as an honor to his old teacher. In 1940, he transferred it to the Naturfredningsforeningen (the Norwegian Conservation Society).

In his deed of gift he wrote:[20]

> "It is my wish with this gift to preserve for all time the renowned mineral occurrences on the island in their present condition. They may therefore not be disturbed. However, the right to take minerals for scientific use is reserved to me and to future directors of the Geological Museum."

The privilege of collecting minerals on Låven was extended to a group of mineralogists, including myself, in an excursion organized by the International Geological Congress in 1960, and led by Dr. Henrich Neumann, the then Curator of the Museum.

After Goldschmidt returned to Norway in 1935 he received a letter from Zachariasen, now Professor of Physics at the University of Chicago, who expressed a keen desire to return to a position in Norway. Goldschmidt encouraged him and actively explored the possibilities, not only in Norway but also in the other Scandinavian countries. For a time there was a possibility of a physics professorship in Bergen, but this came to nothing. It is interesting to speculate on the potential result had Zachariasen returned to Norway at that time, since he played a vital role in the Manhattan Project in the elucidation of the crystal chemistry of the trans–uranium elements.

There was an extensive correspondence between Goldschmidt and Zachariasen between 1936 and 1939. On February 5, 1939 Goldschmidt wrote as follows:

> "I am now host to a 70–year old aunt from Berlin, because of conditions in Germany. I work hard to bring relatives and friends from Central Europa, about half my time goes to correspondence on these matters. . . . Last year, on a bet from four young ladies, I lost 16 kg. Unfortunately, during the winter, because of good Christmas food, I gained most of it back, and have now begun a new reducing treatment. I always feel better when I am a little thinner" (translated from Norwegian).

The Zachariasen family and the Thomassen family visited Norway in the summer of 1939, which pleased Goldschmidt greatly. He was clearly much taken with Zachariasen's young son Frederik. Another visitor that fateful summer was Professor Lark–Horowitz of Purdue University, with an invitation to him to be a guest lecturer at that university in the spring of 1940, another American invitation that was to be nullified by political events. After the outbreak of war on September 1, the Zachariasens and the Thomassens left for the United States on the first available steamer.

When the U.S.S.R. attacked Finland on November 30, 1939, Goldschmidt wrote to his old friend, Professor Pentti Eskola:

> "Dear friend, The dreadful attack on your land and city distresses me greatly, and I am anxious to know of the situation of you and your family. Here in my home we speak daily of you, and I sincerely hope that you have it as well as can be expected under present conditions. I write to remind you that I stand ready to help you in any way. If you consider the possibility of sending your wife and daughter abroad, they will always be very welcome as my guests. If your country's fate should be so cruel that you yourself consider coming to Norway, you are always welcome, both in my home and my institute, and it will be a pleasure and an honor to do anything for you, if there is anything you need that I can send you, just let me know.
>
> I greatly admire the Finnish people's heroic defence against the aggressors, and I only regret that I, an old and pacific person, cannot give more active expression of my feelings and desires.
>
> My housekeeper is also full of concern for you and your family, and also the chauffeur of the hotel limousine asked me yesterday if I had heard from you; both send their warmest regards and best wishes" (translated from Norwegian).

Professor Brøgger, Goldschmidt's esteemed teacher and oldest friend, and one of Norway's greatest scientists, died on February 17, 1940, in his eightyninth year, mercifully spared the spectacle of the German invasion of his beloved homeland. His funeral service in the University Aula (assembly hall) was held in the presence of the King and Crown Prince and a large gathering of scientists and other dignitaries from the Scandinavian countries. At a memorial meeting of the Videnskaps-Akademi on March 8, Goldschmidt spoke at length on Brøgger's contributions to mineralogy and petrology.[21]

His longtime friend, George de Hevesy from Copenhagen, spent a week as his house guest at the end of February and the beginning of March, during which time Hevesy gave an invited lecture at the Videnskaps-Akademi, *Isotopernes Anvendelse i Biologien* (The use of isotopes in biology), a subject for which he received the Nobel Prize in Chemistry for 1943.

Tom. Barth returned to Oslo from the Geophysical Laboratory on April 1, 1940, on the last ship from the United States before the German invasion. On Monday, April 8, it was announced that the British had laid mines in Norwegian territorial waters, to hinder the use of coastal waters by German ships carrying iron ore from Narvik to Germany. The Norwegian parliament was called into special session. In the excitement, few took note of a radio statement that afternoon that a ship had been sunk by a submarine off the south coast of Norway and that drowned horses and uniformed German soldiers were washing ashore. That evening I enjoyed a movie and a late supper with a friend, before taking the last train back to the student hostel at Blindern. As I was walking from the station just after midnight, the air raid sirens sounded. I thought nothing of it at the time (it was customary to test the sirens from time to time) and went to bed, to be awakened at dawn by one of my fellow students and dragged to the window to see German warplanes flying freely over the city.

King Haakon and the Norwegian government left Oslo on the morning of April 9. On the same evening, Vidkun Quisling proclaimed himself chief of state and his Nasjonal Samling (NS) party the ruling political party. This aroused such opposition that the German military authorities dismissed him after a few days. However, thanks to the German Reichscommisar for Norway, Josef Terboven, the NS steadily gained influence with the German civil administration. Opposition grew steadily, and on May 15, 1941, 43 organizations, including the university teachers, sent a letter to Terboven protesting illegal acts of the NS and the occupation authorities. Terboven replied on June 18 by calling representatives of the 43 organizations to a meeting in his headquarters, denouncing them, arresting several on the spot, and declaring some of the organizations dissolved. On September 11, 1941 the elected Rector of the University, Dr. Seip,[22] was dismissed and imprisoned, and Dr. Adolf Hoel,[23] a member of the NS, appointed in his place.

These developments had little immediate effect on Goldschmidt. He considered it his duty to continue to serve the university and the state to the best of his ability. He stepped up the work of the Raw Materials Laboratory, especially the utilization of low-grade phosphate deposits in the Fen area as a source of phosphate fertilizer. He also wrote Part X of his *Geochemische Verteilungsgesetze der Elemente* titled *Über die Verteilung einige Elemente in Meteoriten* (with Aslak Kvalheim as co–author); it was submitted to the Videnskaps-Akademi on May 24, 1940, but never published. He then began work on Part XI; in a letter to Assar Hadding (September 25, 1941) he wrote: "I now have material ready for Part XI of the geochemical series, and I hope that both Parts X and XI will be published within the next year." It was not to be.

In June 1940 his house at Holmenkollen was commandeered by the Germans, and he moved, with his aunt and Miss Brendingen, to an apartment at Holmendammenterrasse 25 in the suburb of Smestad. His aunt died of cancer in November 1940.

Dr. Nils H. Houge has provided an interesting account of Goldschmidt's private life during the critical years of 1941–1942:[24]

> "When you visited Professor Goldschmidt in 1940 he had a house in Holmenkollen. Later on in 1940 he rented the ground floor of a newly built two–family house below Holmenkollen hill. In January 1941 I bought this house from the builder who lived on the second floor, and on the condition that I could move into the ground floor which had access to the garden. I was at that time a medical student, and Professor Goldschmidt was, as one can understand, not too pleased to have to give way, but he agreed to move upstairs, where he installed himself with his housekeeper, Miss Marie Brendigen, and a maid, and where he had three offices for his different activities. I am happy to say that the relationship developed in a positive way. We had a German anti–aircraft battery not far away, and when it was active we all took shelter in the basement, which also brought us closer.
>
> My wife and I had at that time a son, born 1938, and a daughter, born 1941, and Professor Goldschmidt took a great liking to the children. Our son, then three years old, had free access to go upstairs, where Professor Goldschmidt put his work aside and played with him, as by putting chairs in a row on which he placed the boy, the housekeeper and the maid and he himself being the conductor of the "train." We also played "air raids," in which everybody in the flat had to hide under tables while Goldschmidt went around banging blown–up paper bags.
>
> I remember standing in the garden with him and looking at the children at play. He then said: "Dr. Houge, you are a lucky man." I answered that he too, as a famous scientist, had much to be pleased about. His answer was: "I can tell you one thing. Sometimes I regret that I never went out and drank beer with my fellow students." So everything has its price.
>
> When the Jews in February 1942 had to register with the Nazi authorities his aunt, a sweet old lady who lived at Lillehammer, was transported to Germany together with others of the 1500 Jews in Norway. She was never heard of again. Goldschmidt himself

believed that the authorities needed him as a scientist and therefore would not permit his arrest. He told me that he had been called in by the Gestapo, but was set free when he produced a document, signed by Goebbels and one of the other Nazi leaders, and in which he in 1935 was allowed to resign as professor in Gottingen. In spite of this he was arrested in October 1942. Miss Brendigen immediately informed his lawyer, Dagfinn Dahl, now deceased, who mobilized the university and got him free. He was arrested again in November 1942 and was standing on the quay in the second row of Jews to be shipped with the prisoner–ship "Donau." When the first row was marched on board he was called up and released. By now he understood that his luck was running out. A few days later he came to me and said in his correct Norwegian, but with a German accent which he never lost: "Dr. Houge, I have a terrible cough. Do you think that you could give me something for that cough?" I promised that I should get him some tablets. He then added: "I cough quite a lot, so if you could give me plenty of tablets," which I promised him. And not surprisingly he disappeared a few days later. When he returned from England to Norway in 1947 he told me that he would be transported towards Sweden hiding in a waggon of hay, and that he expected there would be children in the consignment who might give them away if they coughed. He told me that there had been about 20 persons, grown–ups and children alike, so his foresight came in useful. When he returned from England in 1947, he was clearly marked by his condition—insufficiency of the heart—which also led to his death which took place here in his home" (original English).

The political situation in Norway changed much for the worse in January 1942. Reichscommisar Terboven instituted an NS government led by Quisling, with Gulbrand Lunde as Minister of Propaganda. All Jews were required to fill out a questionnaire detailing their ancestors back to their grandparents, and their identity cards were stamped with a large "J". Goldschmidt wrote a memorandum, dated March 5, 1942, regarding this:

"In accordance with the decree of January 20 I sent my identity card to the University Secretariat on February 12 with the following declaration. I have no documents regarding the religion of my four grandparents. I have good grounds for believing that at least two of my grandparents were not members of the Jewish religious community, at least in later life. I am not now a member of the Jewish religious community, but from February 1933 to August 1935 I enrolled in the Jewish community in Göttingen; my father enrolled at the same time. My mother was never a member of the Jewish religious community" (translated from Norwegian).

Hans Suess,[25] then at the University of Hamburg, visited Goldschmidt in the spring of 1942, and writes as follows:[26]

"When in 1942 I was sent to Norway for the first time by the German military, supposedly as a consultant for heavy water production, I was anxious to visit Professor Goldschmidt, who for many years had been corresponding with my father in Vienna. I found him planting potatoes with his assistant Dr. Ramberg in front of a large building that housed his Mineralogical Institute. To me he appeared amazingly confident about the future. He said the "Nazi spook" would soon disappear and a new German government would negotiate peace with the allies. I suggested that, in any case, it would be safer for him to go to Sweden and if possible to England."

That he was not really so optimistic can be seen from a story told by Paul Rosbaud.[27] During the occupation of Norway Goldschmidt carried a capsule of poison in his pocket for emergency use. On one occasion a university colleague asked him for a similar capsule. "This poison," he answered, "is for professors of chemistry only. You, as professor of mechanics, will have to use the rope." This was his typical grim humor.

In March 1942 Quisling decreed all Jews in Norway to be illegal immigrants, an ominous pretext for ultimate deportation. As a possible protection, Goldschmidt obtained a statement from his physician, Dr. R. Hatlehol, as follows:

"Professor Dr. V. M. Goldschmidt has since 1939 been under my care for high blood pressure and chronic kidney trouble with the formation of stones. He also suffers from heart disease. He must maintain a strict diet and avoid physical exertion. His condition has required two periods of hospitalization, in 1937 from January 14–27 diagnosed as nephritis chronica, vitium organicum cordis, and again from March 6 to April 11, 1942" (translated from Norwegian).

The protection was illusory. On October 25, two policemen came to his home with an order for his arrest and the confiscation of his property, allowed him to pack some warm clothing, and took him to Bredtvedt prison in Oslo. The following day he was transferred to the Berg concentration camp near Tønsberg, 50 miles south of Oslo. Coincidentally, Gulbrand Lunde, who might have protected Goldschmidt, died in an accident on October 25. He was travelling with his wife in his official Mercedes; it was being loaded on a ferry near Ålesund on the west coast when the craft moved forward and the car and its passengers sank to the bottom of the fjord.

Goldschmidt was clearly a sick man, and the camp physician, Dr. Jervell, had him transferred to the County Hospital in Tønsberg. There he met two other Jewish prisoners, Moses Katz, an orthodox Jew and a hosiery peddler, and Lesser Rosenblum, socialist, atheist, and manufacturer of umbrella handles. Goldschmidt suggested that they should remember the names of their oppressors, so that any survivors could exact retribution. The reply of the pious Moses Katz was a surprise: "Revenge is not for us; that must be left to the Almighty." Goldschmidt asked what prayers would be permissible to God from men in their position. Katz replied without hesitating a moment: "You may pray that the hearts of your enemies may be enlightened." Goldschmidt, unconvinced, turned to the atheist Rosenblum and asked for his view. His reply was equally unexpected: "We must break the evil circle of retribution, or there can never be an end to evil."[27] This experience affected him deeply. Shortly before his death he wrote to Paul Rosbaud:

> "The wisdom of the Moses Katz principles is undeniable. . . . And I am fully convinced that it is my duty towards science and decency to stand firm in continuing my work as long as health permits, thus giving an example to at least some of my junior colleagues. Often I think that (to maintain these principles) to be even more important than my contributions to scientific and industrial research and my scientific teaching. To set a new standard of morality is a matter of great urgency in these times" (original English).

Moses Katz and Lesser Rosenblum were deported to Poland and died in the Auschwitz concentration camp.

Goldschmidt gave the following account of his release:

> "While I was a prisoner in the hospital at Tønsberg, I was visited by my housekeeper, Miss Brendingen, and my secretary, Miss Else Høeg-Omdal, who informed me that the university had begun an action to release me. Miss Brendingen said that a serious point for my release was a demand that I should promise not to leave the country. I replied that I would have to agree to this. But the promise was never demanded of me, either orally or in writing, and I later learned that my release resulted from the actions of the Ministry of Agriculture, and was not linked to any promise from my side. Release was approved November 5, based on my work on Norwegian raw materials, especially the provision of phosphate fertilizer, being of vital importance. I was released on November 7 and returned to Oslo the following day. On November 9 I returned to my work at the university and the Raw Materials Laboratory" (translated from Norwegian).

His release was celebrated at the museum by a small gathering of his colleagues and a toast in cognac, drunk from laboratory beakers. It was no ordinary cognac; the label showed that it came from Roald Amundsen's South Pole expedition of 1910–12!

The reprieve was temporary. Goldschmidt's account continues:

> "Quite unexpectedly a policeman came to my apartment on the morning of November 26 to arrest me. Since I was then at the

museum, he came there and arrested me. I told him that I had been released from Berg concentration camp in order to carry out vital work for the state and showed him my release letter of November 7, but he replied that this had no significance. He took me to my apartment in a taxi, told me to pack a suitcase with necessary clothes, a blanket, food for four days, and family mementos such as photographs. The policeman was at first rather gruff, but later he became polite. I should note that while I was packing I received a telephone call from a lawyer, Helge Skjerve, who said he had been appointed receiver of my property. I told him I was not permitted to speak with him, as the police forbade any telephone conversations.

I was taken to police headquarters at Henrik Ibsens gate 7, where my name was recorded. I was able to telephone the Commerce Department and report the interruption of my work. With other prisoners I was taken by taxi to Bredtvedt prison, where we were registered, our baggage checked, and taken to a new barracks with 14 beds in our room. We were served an evening meal of bread, butter, and tea, and the next morning a similar breakfast. Around 9 a.m. each of us was asked if we were married to a Norwegian woman. I said no, I am unmarried, and we were transported in two small gray busses to the Norwegian–America Line pier, where lay the 9000–ton transport "Donau" from Bremen.

We were lined up in rows of three; I was in the second row. At 10:50 the same policeman who had arrested me called out for Victor Moritz Goldschmidt. When I replied he told me that I was free to go home. He said he was pleased to be able to do this. I told him I had no money to return home, since my money and ration cards had been taken from me when I was registered at the prison. He then requisitioned at police expense a taxi for me. As I was driven from the pier a German soldier asked to see a pass. I had none, but he was satisfied with my "J" stamped identity card. I went home and continued my work at the museum.

On Sunday, November 30 I was visited by the lawyer Skjerve, who said he found it meaningless that my property had been confiscated, and that he would try to have it released. He asked for a written account of my civil service work, and this I sent him on December 1" (translated from Norwegian).

Goldschmidt believed that his release had been engineered by the university authorities, but this may be only partly true. Aslak Kvalheim recalls that when the news of Goldschmidt's impending deportation reached the museum on the morning of November 27, an instrument maker, Ole Nielsen, said "Now it is my turn to do something." Nielsen had been a colleague in the Norwegian Army with a Major Marthinsen, who now headed the Quisling police. Nielsen went to police headquarters and obtained a meeting with Marthinsen, who informed him that, in spite of appeals from the university authorities, no action would be taken to free Goldschmidt. When Nielsen explained the importance of Goldschmidt's work on fertilizers for food production in Norway, Marthinsen became interested and asked whether anyone else could continue this work. Nielsen denied this, whereupon Marthinsen phoned the Agriculture Ministry, evidently confirmed Nielsen's statement, and issued the order for Goldschmidt's release.

After his second release Goldschmidt commented:

> "I continued with my work and tried my best to get my instruments returned, because their confiscation made impossible the work of the Raw Materials Laboratory and much of the instruction of my students. On December 13 the electricity meter in my apartment was read "on account of my bankruptcy," from which I concluded that my apartment would soon be confiscated" (translated from Norwegian).

Hans Suess visited Goldschmidt in mid-December 1942, and gives the following account:[28]

> "When Hans Jensen and I arrived, a table was covered with all kinds of delicatessen, and Goldschmidt said: "Help yourself; this was all confiscated by the Gestapo. Eat as much as you like." A distinguished lady was acting as hostess. A severe anxiety got hold of me, a sinister fear that I had not felt since, for the first time, I had heard of mass murders and genocide about a year before. Jensen and I tried to convince Goldschmidt by all means to leave Norway and escape to neutral Sweden, but he said he did not wish to leave his post as a University Professor and scientist. At the very last minute, when we left, the lady asked: "Do you ever get to Berlin? Then please try to get in touch with Mrs. Furtwängler (the wife of the symphonic conductor Wilhelm Furtwängler) and tell her about the situation." And Goldschmidt said "Wenn ich nun doch in den sauren Apfel beisen sollte. . . ." (If I should now have to bite into the sour apple [to leave Norway], then give my regards to all our friends.)"

On December 18 Goldschmidt was picked up by the Norwegian resistance movement and transported to Sweden. The movement was very active in rescuing Norwegian Jews, and succeeded in sending some 1100 to Sweden; of 760 sent to Poland only 24 survived to return to Norway after the war.

Goldschmidt gave the following description of his last day in Oslo:

> "On December 18 I made a last effort to obtain the release of my scientific equipment. I arranged with Professor Solberg (dean of the science faculty) to meet with Prorector Hoel and inform him that I could not carry out my work at the university because of the

confiscation of my equipment. Prorector Hoel explained that I was in a precarious situation, but that he had an appointment with Statsminister Quisling on December 21 and would take the matter up with him. I became very angry, and said to Hoel that I had the impression of a gangster government, under which one could not carry out one's official duties, was repeatedly arrested, with threats of deportation equivalent to a death sentence. I said that while I had hitherto considered it was my duty to continue my work and it was beneath my dignity to leave the country, I would now reconsider my position, since it was impossible for me to continue my work in Norway. The conference lasted from about 3–4 p.m. and took place in the rector's office; present were Hoel, Solberg, and a secretary, if I remember correctly. I returned home and contacted a friend regarding the possibility of going to Sweden in the near future. He informed me I could leave the same evening. I immediately decided to go. I had already considered the possibility, and I had given my already–written Christmas cards to Mr. Granli at the museum, to be mailed at the appropriate time" (translated from Norwegian).

In a letter to Ole Nielsen (October 30, 1945) Goldschmidt recalled his escape from Norway:

> "I have often dreamed of being brought to the slave ship Donau at the Amerika Line pier and standing in line to be marched on board. and I still dream of one moonlight December night tramping through a birch forest over the frontier to Sweden and being warmly welcomed by soldiers at a small barracks, together with almost 40 fellow escapees, among them a boy only 3 years old who was travelling alone to join his father. And the next morning, breakfasting on white bread, butter, and cocoa, and enjoying a sauna bath. I am deeply grateful to those who aided my escape from Norway, the people who guided us over the border, carried an amputee on a stretcher, and helped old ladies and young children" (translated from Norwegian).

On January 5, 1943, Rector Hoel learned of Goldschmidt's escape to Sweden from Professor Solberg, who had received a telephone call from Miss Brendingen, who said that on December 18 Goldschmidt had packed a bag and left home, since when she had heard nothing from him. Hoel decided that he was obliged to inform police chief Marthinsen, whom, he reported, was appalled and exclaimed "Yes, I should have known better than to believe you blue–eyed idiots" [Norwegian slang for childish naiveté]. Hoel was interrogated by the police and by the German Sicherheitsdienst, but no further action was taken against him.

Conditions at the university reached a crisis in late 1943. The commencement of the 1943–44 academic year had been postponed until October 1, so that students could assist with harvesting. In September the government Department of Education, now in the hands of Quisling appointees, issued new regulations for university entrance, one of which provided for 25% of students being admitted without reference to academic qualifications–clearly an effort to "nazify" the student body. The university authorities protested strenuously, and began negotiations with the government authorities. These negotiations were cancelled by Quisling on October 14, and the following day 10 professors and 50–60 students were arrested. Worse was to come. On November 28, there was an arson attempt at the university. It caused little damage, and was probably an NS provocation—a Norwegian Reichstag fire. On the evening of November 29, the underground learned that at 11 a.m. the following day the university was to be occupied by SS troops, and all male students arrested and deported to Germany. Many students were forewarned and went "underground" or escaped to Sweden, but about 700 were sent to "reeducation camps" in Germany.

The research institutes at the university were able to reopen after a short time, and provided a cover for continued underground activities against the German occupiers and their Norwegian collaborators. Arrests continued from time to time, and many students and professors ended up in German concentration camps, among them Goldschmidt's colleague Anatol Heintz (Professor of Paleontology).

On February 8, 1945 police chief Marthinsen was assassinated outside the university at Blindern. In reprisal, 37 Norwegian hostages were executed. The war in Norway ended with the surrender of the German forces on May 8, 1945. Reichscommissar Terboven committed suicide. Quisling was arrested on May 9, tried for treason, found guilty, and executed on October 24, 1945.

In 1946, Adolf Hoel was the subject of a judicial inquiry into his position in the Quisling NS party and his actions during the war. Goldschmidt wrote the following letter in his defense:[29]

> "On several occasions Adolf Hoel supported me when my life was in danger during the persecution of the Jews in Norway in 1942. I know that he participated in the action to release me from the Berg concentration camp, and he probably guaranteed me when I was released in mid–November in order to continue my researches on Norwegian phosphate deposits. The state police, however, disavowed this agreement, and arrested me again at the end of November with the intention of deporting me on the prison ship "Donau" the following day. Immediately before the ship's departure I was released, probably because of protests from Adolf Hoel and others to "Major General" Marthinsen.
>
> On December 18, 1942, I met with Adolf Hoel to protest the confiscation of my apartment, my instruments, and my library, since without them I would be unable to continue my scientific

research. I told him that under these circumstances I could no longer remain in my position, since I could no longer serve my country. Adolf Hoel could not promise that I would be able to recover my scientific equipment, but promised to take the matter up with the appropriate authorities.

The same evening I learned that I could join an illegal fugitive transport to the Swedish frontier. Since I felt that because of certain Allied interests my departure was highly desirable, I accepted the opportunity and crossed the border to Sweden the same night. On arrival in Stockholm I reported to the Norwegian authorities, and continued as soon as possible to the United Kingdom, although delayed by an unforseen hospitalization. I arrived in the U.K. on March 3, 1943.

I would emphasize that Adolf Hoel was never aggressive against me, although he was certainly aware of my detestation of Naziism, a detestation which I have never concealed, either in this country or when I lived in Germany from 1929 to 1935.

As a scientist I deeply deplore that Adolf Hoel was so deluded as to join a party that was to become a disaster for our country. I must assume that he intended to serve the country. In the autumn of 1942 he volunteered to me that he considered it his responsibility to try to bring Norway through the war with the least possible damage.

Concerning Adolf Hoel's appointment as "prorector," I consider that, of the possible alternatives, it was an advantage for the university to have as "prorector" an able scientist with an understanding for practical administration. I believe that with a bureaucrat or politician as a "Nazi prorector" there would have been much more serious conflicts than with a scientist like Adolf Hoel who had a practical grasp on the realities.

When I was a committeeman in the Scientific Association in Göttingen (around 1930—I do not recall the exact date) I was glad to cooperate in honoring him by his appointment as foreign corresponding member of the association, as a recognition as the foremost Norwegian Arctic scientist after the death of Fridjof Nansen" (translated from Norwegian).

9 / Last Days—Sweden, Great Britain, Norway: 1942–1947

AFTER GOLDSCHMIDT ARRIVED in Stockholm on December 19, 1942 he was joined by the "distinguished lady" mentioned as his hostess by Hans Suess. She was Mrs. Sigrid Elena Grüner–Hegge (née Feinsilber), and had separated from her husband, who was Director of the Filharmonisk Selskaps (Philharmonic Society) orchestra in Oslo. I recall her as a very attractive and vivacious woman, about 30 years old. Goldschmidt introduced her as his fiancée, and proudly announced "Now I am going to get married and have a lot of Jewish children—that will make Hitler mad!" Goldschmidt's base while he was in Stockholm was the Mineralogical Institute of the university, so we saw quite a lot of him and his lady; my fellow students and I were naturally quite intrigued with the situation, since this appeared to be his first romantic association.

It was clearly a great relief to Goldschmidt to be in a neutral country and to enjoy the hospitality of his many friends. The university offered him a position as guest lecturer, and he planned a series of ten lectures describing his geochemical researches, to be given during the month of February, 1943. Only three were actually given; after the first lecture on February 3 he suffered an internal bleeding and was hospitalized in the Ersta Hospital for three weeks. He gave the second lecture on February 26 and the third on March 1. The series was terminated when he was flown from Sweden by the British authorities on March 3.

Goldschmidt could have remained in Sweden. The current holder of the chair of mineralogy at Uppsala University, Dr. Helge Backlund[1] had reached retirement age and the position was offered to him. However, he felt strongly that his duty lay with the Norwegian state, and that his knowledge of technical developments in Norway and the German interest therein would be of great value to the Allied cause.

Goldschmidt landed at a military airfield at Dyce, near Aberdeen (where I was to arrive the following August), at about 10:30 p.m. on

March 3. The next evening he was taken by train to London, thence to a school in south London, where he spent several days being debriefed; he commented that the school was very cold and drafty. After he was released, the Norwegian authorities arranged accomodation for him at the South Kensington Hotel, where he lived until he left for Aberdeen in August.

Goldschmidt's romance with Mrs. Grüner–Hegge eventually expired. She followed him to England and settled in London, where her brother, Dr. Johann Feinsilber, had a dental practice. They evidently saw rather little of each other after he moved to Aberdeen in August 1943. On November 5, 1943, he wrote to Percy Quensel, Professor of Mineralogy at the University of Stockholm:

> . . . . "Sis [her familiar name, BM] has got some work in London, I saw her only for some few days in August. Her application for a divorce cannot be granted before next summer, according to present Norwegian regulations. My family plans thus have to be postponed, a very great disappointment indeed. I have to thank you and other friends very much for all your kindness towards her" (translated from Norwegian).

When he received the Wollaston Medal from the Geological Society in 1944 he asked that she be invited to the ceremony, and wrote to Professor E. B. Bailey (March 3, 1944):

> . . . . "I have to thank you also very much for your great kindness to arrange the invitation of Sigrid Elena Grüner–Hegge to the lunch on March 15. I owe my escape and my life to her courage and determination."

Dr. Kathleen Lonsdale wrote to Goldschmidt on April 11, 1944:

> "Mrs. Grüner–Hegge had lunch with me at the Royal Institution today and I was glad to know that she hopes to go and stay with Mrs. Born later on. I am sure that she needs friendship very much. I was able to convince her that the Duke of Bedford's concern about the morality of regulation 18B,[2] a concern which is shared by all Friends [Quaker], does not mean that he—or any Friend—has any sympathy with the views of Mosley or the Fascists. If it can be arranged, I think she will now be willing to go and stay at his home after she has spent a little while with Mrs. Born. The friendliness of a large country house will be very refreshing to her, I think, but I do feel at the same time that she may, in the long run, be happier with her own people in Sweden than she is here."

Goldschmidt replied on April 15:

> "I have to thank you very much for your kind letter of April 11 and for all your kindness and helpfulness towards Mrs. Grüner–Hegge. I quite agree with you that she needs friendship and some

time of quiet and rest in a friendly atmosphere and I am very glad to learn that she is considering such an idea. Please let me know, if any financial assistance from my side could be helpful, for instance to cover the costs of her travelling. I am very much indebted to her for her most active help in making possible my escape from Norway, when she risked her own life to save a comparative stranger, as I then was. For that same reason I am not very inclined to recommend her return to Sweden during the war. In case Sweden is overrun by the Germans, Mrs. Grüner–Hegge may risk severe reprisals from the Germans for her assistance to me in 1942, when I left Norway."

By 1945 the romance was evidently over. In a letter to Dr. Felix Pincus, Monroe, Michigan, on February 2, 1945, he wrote:

> "I have, for the moment, given up the idea of marrying. Or rather the lady inferred did it, as we did disagree in nearly all vital points and interests, which really again may have been one of the close shaves, presented to me by a friendly fate."

Mrs. Hedwig Born (wife of Professor Max Born, friends of Goldschmidt since the Göttingen days) may have expressed the appropriate epitaph when she wrote to him on July 18, 1946:

> "I wonder whether you ever hear what role Sis is playing now. Perhaps she is back with her husband after all. I'm sure you can hear of her now without any stir of your heart! You had a happy escape, but on the other hand you owe her that you are now back in Oslo and not buried in Poland. In Sis I have a novelist's interest, she was the first of her species I came to know. Of course she has splendid qualities, nobody can deny that."

In 1953 Mrs. Grüner–Hegge married Aake Anker Ording, a Norwegian civil servant who also worked for the United Nations. She died in 1977.

Goldschmidt's months in London were extremely busy ones; in a memorandum dated July 26, 1943, he lists some 150 conferences in which he took part. Many of these were with Norwegian and British authorities on conditions in Norway, especially in regard to the German exploitation of raw materials and the production of heavy water; the latter was of particular interest because of its possible use in the fabrication of an atomic bomb. He attended conferences at the University of Cambridge, in Manchester, Sheffield, Edinburgh, and Aberdeen. On May 8, he lectured at the British Coal Utilization Research Association on the occurrence of rare elements in coal ashes. He was also concerned with his own business affairs, especially the utilization of his many patents on ceramics and refractories. His United States patents, from which he had received considerable royalties from the Harbison–Walker Refractories Company of Pittsburgh during the past ten years, had been sequestered by the Alien Property Custodian, and he tried (without success until 1945) to have them returned to him.

There were some problems *vis-à-vis* his official position with the Norwegian government. He maintained he was still a university professor and should be paid accordingly. However, the Prime Minister of the government-in-exile, Johan Nygaardsvold, maintained that he had come to England at the request of the British authorities, not of the Norwegian government. The situation was resolved by the good offices of the British Agricultural Research Council; a letter dated August 7, 1943, from the secretary of the council reads as follows:

> "I am directed by the Agricultural Research Council to refer to informal consultations with Professor V. M. Goldschmidt, F.R.S., and Lieut. Fürst, and to seek the consent of the Royal Norwegian High Command to an arrangement whereby, for a period of six months in the first instance, Professor Goldschmidt would be permitted to devote about three-quarters of his time to soil investigations on behalf of the Agricultural Research Council at the Macaulay Soil Research Institute, Aberdeen, and at other centres. For their part the A.R.C. would be prepared to refund to the Royal Norwegian High Command three-quarters of Professor Goldschmidt's salary, which, it is understood, is approximately at the rate of £112 per month, and to meet his travelling or other special expenses in connection with the researches on which he would be engaged for them."

There is some evidence that this arrangement was devised to remove him from London and rusticate him in far-away Aberdeen. He apparently caused some concern by talking too freely about atomic energy developments. The possibility of making an atomic bomb, highly classified in the Allied countries, was freely discussed in scientific circles in Sweden, and articles on the subject had appeared in Swedish journals. When I mentioned this to scientific friends in England after my arrival there in 1943 I was told on no account to share my knowledge, little though it was, with anybody—the penalty might well have been internment.

Goldschmidt arrived at the Macaulay Institute on August 26. He was suffering from a recurrence of his medical problems, and spent six weeks in treatment at the Royal Infirmary. After his release, he worked intensively at the Macaulay Institute, partly on soil problems, but at the urging of the Director, Dr. W. G. Ogg, he devoted much of his time to writing his *magnum opus,* a comprehensive treatise on geochemistry. He also continued to travel extensively, giving invited lectures.

During this time he also received numerous honors. In May 1943 he was elected a Foreign Member of the Royal Society, and in January 1944 he was awarded the Wollaston Medal, the highest honor of the Geological Society of London. He particularly prized these awards, since they had also been given to his teacher and friend, Professor Brøgger. At the presentation of the Wollaston Medal on March 15, 1944, the President of the Geological Society, Professor W. G. Fearnsides, spoke as follows:[3]

> Professor GOLDSCHMIDT: The Council of the Geological Society have this year awarded to you the Wollaston Medal, the highest token of appreciation it is within their power to bestow. The fields of your accomplishment are wide; they are concerned as much with crystal chemistry and geochemistry as with field petrology, and in each of them you have made outstanding contributions. Born in Zürich, you went as a boy to Kristiania, and through your University course and afterwards had the good fortune to work with that greatest of all Norwegian geologists, Professor W. C. Brøgger. By your 1911 monograph on contact metamorphism in the Kristiania region, you laid the foundations of systematic study of the mineral transformations which proceed under changing conditions of earth temperature and pressure. Your descriptions of the aureoles around the intrusive igneous complexes of that region are the first successful essays towards a systematic classification of rock–mineral associations in the light of the phase rule. Your investigations of the zonal metamorphism of igneous complexes among the crystalline schists of the Palaeozoic mountains of South Norway, completed during the following ten years, are not less important. What you discovered in the course of those genetic studies of rocks which had been foliated and recrystallized at high temperatures during the folding has inspired much recent work in the Scottish Highlands.
>
> In 1917 you organized for the Norwegian Government a research committee to promote the use of the country's minerals as industrial raw materials, a service with which you maintained connexion during the following twenty–five years.
>
> Since 1919 you have laboured unceasingly in laboratories, first at Oslo University, and then in geochemical research schools of your own founding, at Göttingen from 1929 to 1935, and then again at Oslo. As methods of analysis you have developed and applied quantitative optical and X–ray spectroscopy, and through a series of elaborate investigations of the chemical composition of rocks and minerals have revolutionized our knowledge of the distribution of the minor constituents of the crust of the earth.
>
> The importance of your classic researches in measuring the ionic radii of most of the elements, and determining the crystal structure of long series of ionic compounds, was recognized by W. H. and W. L. Bragg, two of the most distinguished of British physicists, in the phrase, "the crystal chemistry of inorganic compounds has been built on the foundations which he laid". To us geologists also, your story of ionic structure as primarily determined

by the geometric packing of ions of those sizes which will fit, is fascinating. We have been thrilled by your demonstration that ionic charge and size can determine the partition of the rarer elements among the common elements at each successive stage of the geological cycle. You have made it clear to us that, as in igneous rocks, the rare earths and others of the less abundant elements entering the growing crystals have sieve-sorted themselves according to ionic size and bonding force, so in weathering and through all the processes of sedimentation the distribution of trace elements was ordered according to ionic potential, the ratio of ionic charge to ionic radius leaving the balance of the most soluble ionic compounds in the waters of the sea. Your attribution of the unexpected concentrations of such elements as boron and germanium, arsenic and bismuth in coal and humus, to biochemical processes is of much interest. Your comparative studies of the distribution of elements in furnace slags and meteorites and crustal rocks have brought us an entirely new conception of the earth's interior, and we have appreciated that in formulating principles of geochemical evolution of terrestrial matter you have made contributions of outstanding merit to the understanding of the mineral structure of the earth.

As a protest against the racial policy of the rulers of Germany you resigned the Professorship at Göttingen and, in August 1935, returned to Oslo where also you incurred the wrath of the Nazi invader. After a term of imprisonment and the concentration camp, and the confiscation of all your property, you were sentenced to immediate deportation to Poland.

You had been elected a Foreign Correspondent of the Society in 1923, and a Foreign Fellow in 1931, and we rejoiced to have you here in March last year. You have suffered to uphold the cause of knowledge and of freedom, and have well served the science of the earth. We salute you as a petrologist and geochemist and also as a distinguished representative of an allied nation.

Professor GOLDSCHMIDT replied:—

Mr. PRESIDENT: First of all, I wish to give expression to my feelings of appreciation for the highest award in the science of geology. In awarding me the Wollaston Medal, the Council of the Geological Society has placed me near my old teacher, colleague and friend, Waldemar Christopher Brøgger, who was not only the greatest geologist of Norway, but also one of the most righteous men I have met.

I consider the award to be an honour also to the five teachers to whom I owe my education in science—besides Brøgger, they were my father Heinrich Jacob Goldschmidt, the physico-chemist, Hans Henrik Reusch, the geologist, Thorstein Hallager Hiortdahl, who taught me crystallography and mineral analysis, and Friedrich Becke, at whose Institute in Vienna I studied petrographic optics one autumn and winter. Both Brøgger and Becke were Wollaston Medallists, and, looking through the list of the awards of the Wollaston Medal, I find also my old friends Paul von Groth, Gerard de Geer, Albert Heim, and Alexander Evgenovich Fersman, the foremost organizer of geochemistry in the Soviet Union. In that great country, through the work of my friends Fersman, Vernadsky and other great scientists, the practical importance of geochemistry was recognized so early by the authorities that the Soviet Union could develop her industrial strength with the full weight of her powerful resources.

From my father and from my other teachers, and from many of my friends, I have not only received much scientific information and advice, invaluable as that has been to me, but they set me examples of still greater importance—never to consent, not even tacitly, to activities contrary to their fundamental ideas of the importance and dignity of science. Several times in my life—as I hope in faithful obedience to such principles—I found it my duty to resign my academic position as an ultimate protest against detrimental developments which I was unable to prevent; three times in my life, therefore, I had to organize and equip new laboratories, for a great part at my own expense, and to train new staffs for research in geochemistry. Speaking of the laboratories, I wish to express my gratitude and friendship to my assistants, research workers and students from a score of different nations, throughout thirty-three years.

Work on chemical geology and mineral chemistry is not a novel science in Norway. Besides the great Brøgger, we appreciate early work by his predecessor Theodor Kjerulf, by the mineral chemist Teodor Scheerer, as well as in more recent times by Johan Herman Lie Vogt, professor of metallurgy, whose application of physical chemistry to igneous rocks earned him the Wollaston Medal. Even in the first half of the nineteenth century, Kjerulf's predecessor, Balthazar Mathias Keilhau, the author of admirable geological maps, wondered about problems of metamorphosis and granitization, but paid little attention to chemical evidence, as do even some modern representatives of that line. Among the younger and youngest generations of Norwegian geologists, too, many lines of chemical geology are represented.

In the award of the Wollaston Medal to me, at an age decidedly below the average age of the previous medallists, I see encouragement for continuing my activities in that branch of science which I have tried to develop during the last quarter of a century, namely, modern geochemistry, based upon atomic physics and atomic chemistry.

It is a branch of science needing careful organization and, for experimental work, considerable means—means, much of which I have hitherto been able to earn through my own work in applied science. The means for that work, however, were for a great part lost when, in 1942, Hitler and Quisling confiscated all my property in Germany and Norway, with such careful efficiency that not even the watch in my pocket or the wedding rings of my late parents were left to me. However, I feel fully confident that for a fourth time I shall be able to reorganize my research work on geochemistry.

One may wonder whether the new geochemistry is still a branch of geology or if it belongs rather to chemistry or even physics. I think it serves no useful purpose to construct strict lines of division, cutting through the unity of modern science. This prominent audience of geologists will agree with me that geology always has been, and still is, in a central position in the general development of pure and applied science. A few great and most important problems are the ultimate objects of modern scientific research. One of these great problems, the evolution of organisms, has been, and is, most intimately connected with geology through the work of your great Charles Darwin, who, in 1859, added new splendour to the Wollaston Medal.

Another fundamental problem, the composition, structure and distribution of matter, an object for modern geochemistry, is as closely connected with geology, even if it demands a kind of "combined operations" of geology with chemistry, physics and other branches of science. Already, too, we can visualize the work of geochemistry as being closely connected with astrophysics and nuclear physics, leading up to the final problem of the origin and evolution of matter itself.

After my experiences of the past years, and my escape from Norway, the generous hospitality of Swedish and British friends and British authorities has made it possible for me to recover my health, and such an award as the Wollaston Medal gives new strength and courage to my intention to continue a line of work which opens new aims and activities for pure and applied geology, as it has done for chemistry and physics.

Most sincere thanks I owe to Norway, where I had the privilege to live part of my boyhood, my student days and nearly all of my later life, in a free country. I wish especially to thank the Royal Norwegian Ministry of Commerce, which always, through a quarter of a century, has had understanding and generosity for my work, and which in 1935, on my return from Göttingen, anew gave me hospitality.

And, most respectfully, I bow my head for the fighting and suffering men and women of the Norwegian people.

In February 1944 the University of Aberdeen awarded him the honorary degree of Doctor of Laws, which was conferred on him at the annual graduation ceremony on June 29 (Plate 41). In 1945 he was elected an honorary member of the Chemical Society (of London).

In 1944 Dr. Ogg was appointed director of the Rothamsted Experimental Station at Harpenden, some 30 miles north of London, and he invited Goldschmidt to join him there at the expiration of his year's appointment at the Macaulay Institute. In July Goldschmidt was in Edinburgh conferring with his old friend Professor Max Born on the structure of water when he suffered a minor relapse and spent some time in a nursing home. Dr. Ogg wrote urging him to postpone his departure to Harpenden. Goldschmidt replied on July 21:

> . . . . "I have to thank you very much for your kind letter, suggesting a postponement of my journey. Please, allow me to disobey your most friendly urge. I have recovered excellently and I am now in a much better condition than for months. . . . I have not come to Britain for the purpose of resting in a nice nursing home, but, if possible, to do some useful work. And I hate the idea to be disencouraged by any considerations to my health, which anyhow, now is quite satisfactory and up to any requirements."

So Goldschmidt proceeded to Harpenden and became a guest in the Ogg home. On the night of December 14, as Dr. Ogg was going to bed, he heard groans from Goldschmidt's room and found that he had suffered a severe heart attack while peacefully reading in bed. A doctor was rushed in and kept him alive with oxygen and injections. He then spent several weeks in hospital before moving to Miss Debenham's Nursing Home, which was to be his residence for his remaining time in England. He recovered sufficiently to be able to work on his book and manuscripts, and even to travel to London occasionally, but his doctor insisted on his living in the nursing home, where emergency treatment was available in case of another heart attack.

With the end of the war in Europe and the liberation of Norway in May 1945, Goldschmidt anxiously awaited news from Oslo. He wrote to Lester Strock in the United States on July 11:

> "I have the pleasant news, that all my instruments, scientific library, and notes have been rescued to the museum, but most of my other belongings, with some few exceptions have been looted irretrievably in the great pogrom, with the exception of some things which Miss Brendingen managed to get aside. I had the very good news indeed, that all my assistants at the Raw Materials Laboratory are alive and tolerably well. Some few, who have been imprisoned in Germany, are still ill, but recovering. My assistants at that laboratory have behaved most courageously, and are doing excellent work the whole time. Also from the Ministry of Commerce I received most friendly and encouraging letters. From the University authorities and from the University people at the museum I have not received a single word yet, nor any reply to my letters. Mr. Kvalheim, in my absence was made deputy director of the Raw Materials Laboratory, by the

> Ministry of Commerce, the most excellent choice, I think. From Stenvik and Kvalheim I had some very nice letters, and from some few other friends in Norway'' (original English).

Among the letters he received was one from his old friend George de Hevesy, who escaped with his family from Denmark to Sweden in 1943 and was working at the Research Institute for Organic Chemistry and Biochemistry of the University of Stockholm (dated May 19, 1945):

> ''Dear Friend Goldschmidt, what a pleasure to hear from you! I was much worried to hear recently that you were in indifferent health and it is with greatest pleasure I am reading that you are in full vigour again and engaged in writing your Geochemistry to which I—and many others—are looking forward with so much pleasure to read. I was deeply touched that you sent me and my wife your congratulations at the occasion of the liberation of Denmark. . . . I fully understand that Ogg wishes you to stay at Rothamsted after the war, but I very seriously hope, in the interest of Norway and of Scandinavia, that you will return to Oslo. How could Norway be without you!'' (original English).

Hevesy's hopes for Goldschmidt were evidently not shared by a considerable faction within the University of Oslo, among them Professor Barth. When Goldschmidt did hear from the University authorities, it was an evasive letter that failed to answer whether or not he retained his professorship. He was given leave from the university until March 31, 1946, to complete his scientific and technical commitments in Great Britain.

Worse was to come. On February 26, 1946, he met with Professors Halvor Solberg[4] and Egil Hylleraas[5] from the University of Oslo at his club, the Society of Visiting Scientists, in London. To his surprise, Professor Solberg (Dean of the science faculty), in the presence of Norwegian and British scientists, upbraided him as a deserter, in that he had left Norway in December, 1942, without informing him and Rector Hoel, thereby endangering their lives as possible hostages. In March, he received an official document from the university with the following accusations:

1. He left Norway to save his own life, and thereby endangered the lives of his colleagues.
2. He had reported Professor Barth's relationship with the German Nazi, Dr. Noack, to the Allied authorities in London, thereby injuring his reputation as a loyal Norwegian.
3. In a private letter to a Swedish colleague he had compared Barth unfavorably to Gulbrand Lunde, the Quisling Minister of Propaganda.

Since no action was taken against either Solberg or Hoel after Goldschmidt's escape to Sweden, the first accusation can perhaps be dismissed. The second accusation was probably directed at the wrong person. The relationship between Barth and Noack was reported to the British authorities by Dr. Felix Singer, who acted as Goldschmidt's patent agent in England and was staying with him in Oslo in May 1939, at the time of the proposed sub–letting of Barth's apartment to Dr. Noack. The third accusation concerns a letter Goldschmidt wrote to Professor Hans Ahlmann of the University of Stockholm on September 16, 1945, in which he compared Barth to Lunde unfavorably "both as a man and a scientist." It is not clear how this letter reached the university authorities, whether it was forwarded by Ahlmann or in some other way. In his reply to this accusation Goldschmidt withdrew his comparison of Barth and Lunde claiming, naively, that the comparison was not intended to relate to their politics.

When Goldschmidt returned to Norway he considered taking legal action against his accusers for defamation of character, but was dissuaded by his lawyer, Mr. Gunnar Mellbye.

Goldschmidt was determined to return to Norway, despite attractive offers of positions in Britain, an invitation to return to his professorship in Göttingen, and a request from the Chinese government for him to organize raw materials research in that country. Continued ill health delayed his return. On February 1, 1946 he was able to regain his apartment at Holmendammenterrasse 25; Miss Brendingen moved in and was able to recover much of his furniture. He had booked a flight to Oslo at the beginning of April, but was not fit enough to travel. He finally flew back on June 26. Kvalheim described his return:

> "He came by air to Gardermoen, direct from his sick bed in England. Friends who had come to meet him hardly recognised him. They remembered him as a robust and hearty person, but now he was only a shadow of his former self. He was stooped, hollow–cheeked, and had dark circles under deeply sunken eyes. His friends saw clearly that he was a very sick man. But he retained his old–time "gallows humour." He greeted his friends by saying "Dont be upset if I become dark blue or black in the face, just give me one of these pills" (translated from Norwegian).

Immediately after his return to Oslo he was hospitalized for three weeks, but then returned to his positions as Director of the Geological Museum and of the Raw Materials Laboratory. Professor Barth had accepted a professorship at the University of Chicago and left for the United States in August; he remained there until 1949, when he returned to Norway to assume the directorship of the museum. Goldschmidt was warmly welcomed back by the staff of the Raw Materials Laboratory and some (but not all) of the museum staff.

During the summer his health improved somewhat, probably due in part to the ministrations of Miss Brendingen, who was an excellent cook and provided him with what he regarded as a gourmet dish, grilled whale steak! Among his visitors was his distinguished research student and old friend Professor Zachariasen. When the university year began in the autumn he lectured regularly to his students. However, the improvement in his health was only temporary. In October he noted a black spot on his leg; it proved to be malignant and required surgery. The carcinoma could not be completely removed, and further operations were performed. Nevertheless, he continued working on his book, directing scientific and industrial research, and corresponding with friends, especially Paul Rosbaud, who had moved to England from Germany after the war. In February 1947, he was invited to be a guest of the Chemical Society at an International Chemical Conference to be held in London July 17–24, 1947, and to be Chairman of Section 1 (Inorganic and Geochemistry). He accepted, and proposed to give two lectures, "The Principles of Modern Geochemistry" to Section 1, and "Forsterite and Olivine as Refractories" to Section 11.

It was not to be. On March 15 he wrote what may have been his last letter, appropriately enough to his long–time friend Paul Rosbaud:

> "I have to thank you very much for your kind letter of March 7th, just with a few lines, as I am in some hurry because of an impending operation again. Many thanks for having sent me the excellent photostat copy of Jensens new publication in *Naturwissenschaften.* I am hard at work on the *Geochemistry.* My heart has been very much improved by the euphillin treatment, which now can be effectuated by os instead by anus or by intravenous injection. But my Nr 1 cancer leg has been disimproving recently. I shall have my sixth operation this winter, to cauterize the affected region under narcosis. Well, I think the heart will stand up to it, I have found it really shock proof this winter, as compared to its resistivity just a year ago, which was not quite sufficient. Anyhow, the wise principles of M.K. [Morris Katz] are guiding and protecting me. So I am going to the Radium Hospital on Monday, after having given my weekly Mineralogy lectures Monday morning. I hope you and your family are quite well. Please note your expenses for me on scientific literature, so that I can regulate matters.
>
> Best greetings from Miss Brendingen and also from Miss Jenny, whom I occasionally see.
>
> Here we have had a very frosty winter, the cold being nearly continually about 20 centidegrees below [−20°C]. And nearly no snow. Recently the ice along the South Coast has been broken by a Northerly gale so that supplies again reach Oslo by sea. The fauna has been much enriched by wolfs which are considered to be Russian ones, even if nobody has spoken with them nor examined their passports, if they had any. One of them proceeded to Denmark on the Kattegat ice. I'm wondering whether we can expect wolf cutlets at Messrs. Jensens or Lorentzens.[6] Bear anyhow is not too bad (if it isn't an old and hoary individual), but silver

> foxes (which I have eaten in Sweden) can not be much recommended.
>
> Well, I am glad to get operated now, because the present condition of the number one leg is rather unpleasant and the pain keeps me awake at night. The number two leg behaves very nicely, indeed" (original English).

The operation was a minor one and he returned home on March 20. Later that day he complained of a sudden intense pain in his head and died almost immediately of a cerebral hemorrhage.

As he had arranged, his body was cremated; his ashes were placed in a sealed metal envelope and deposited in the green olivine urn he had provided. The urn with the ashes was placed in the columbarium of the Western Crematorium in Oslo, but owing to space restrictions, his ashes, together with those of his father and mother, were buried in May 1986 in the graveyard on which the crematorium is located. The urn is preserved at the Geological Museum, together with those of his parents (Plate 43).

10 / *Epilogue*

AFTER GOLDSCHMIDT'S DEATH, Dr. Ivar Oftedal, who was then senior curator, was appointed Acting Director of the Museum, and Aslak Kvalheim was appointed Director of the Raw Materials Laboratory. Tom. Barth returned from Chicago to his professorship in March 1949, and was then appointed Director of the Museum; Oftedal was appointed Professor of Mineralogy after Goldschmidt and moved to the Blindern campus of the university. The Raw Materials Laboratory moved to larger quarters in Trondheim in 1952, and was incorporated into the Norwegian Geological Survey as the Geochemistry Division in 1962.

Kristoffer Stenvik, who had worked closely with Goldschmidt in the development of olivine as a refractory material, continued this work in the Raw Materials Laboratory. In 1947 the commercial exploitation of Norwegian olivine was handed over to a government company, A/S Olivin, and Stenvik worked for that company until 1957, when he returned to the Raw Materials Laboratory. Olivine is now a major mineral product in Norway, estimated production in 1991 being approximately 3 million tons, valued at 35–150 kroner/ton, depending upon quality and degree of processing.

I was able to revisit Norway in the summer of 1948. At the museum I found my room exactly as I had left it more than eight years earlier, even to the spectrographic plates I had developed the day before I left. Dr. Ivar Oftedal was working through Goldschmidt's papers, which included hundreds of pages of manuscript intended for his comprehensive treatise on geochemistry. The outline of the book was there, but some of the chapters were barely started or incomplete, and much of the material was in his handwriting, which was not easy to decipher. We were both very dubious as to the possibility of completing a publishable manuscript. It is to the great credit of Dr. Alex Muir of the Rothamsted Experimental Station that he undertook this task, resulting in the publication in 1954 of Goldschmidt's *Geochemistry*, a 730-page volume. The book reflects the amazing versatility of this great scientist, and his ability to synthesize and correlate a vast amount of scientific data. Goldschmidt's will directed that the royalties should accrue to the Raw Materials Laboratory, which shared them with Dr. Muir.

Following World War II the science of geochemistry enjoyed a vast expansion. Much of this resulted from the enormous demand during the war for raw materials, and from the atomic bomb program, with its emphasis on the geochemistry not only of uranium and thorium but also of the many rare elements produced by their fission. New techniques and instruments for chemical analysis revolutionized the determination of small amounts of minor and trace elements. New fields of investigation, such as the origin of the elements, were opened up. An international journal, *Geochimica et Cosmochimica Acta,* began publication in 1950, largely through Paul Rosbaud's efforts;[1] volume 1 contained 338 pages, volume 54 (1990) contained 3608 pages. The Geochemical Society, founded in 1955, now has some 1800 members in 44 countries; its highest award is the Goldschmidt Medal, presented annually together with an honorarium endowed by the late Dr. Lester Strock.

In 1947 I left New Zealand and joined the staff of Indiana University as Professor of Mineralogy. In addition to my mineralogy courses, I was asked to give a graduate course on a subject of my choosing, and I chose geochemistry. In 1950, I was approached by a publisher's representative. "I hear you're teaching a course in geochemistry" he said, "Would you write a book for us?" My concept of book writing was that the prospective author wore his feet down to the ankles trudging from publisher to publisher while peddling his manuscript—this was an entirely unexpected approach. I replied "If I write it, who will buy it?"—knowing that few if any universities were offering such a course, at least under the name geochemistry. He replied "That's our problem—you write it, we sell it". As a result I wrote *Principles of Geochemistry.* Published in 1952, I believe it was the first textbook of geochemistry in the English language.[2] Geochemistry is now an accepted subject in most universities and many colleges. The 1990 *Directory of Geoscience Departments in the United States and Canada* lists a total of 10,006 faculty members, of whom 1000 list their principal speciality as geochemistry.

In 1953 I accepted the position of Curator of Geology and Mineralogy at the American Museum of Natural History in New York. On my arrival I was interviewed by the then Director, Dr. Albert Parr (1900–1991). He was an expatriate Norwegian, and on learning I had been a research student with Goldschmidt he reminisced on his student days at the University of Oslo. He had taken the introductory course in mineralogy in 1926 as part of the requirements for a science degree, but was not seriously interested and did poorly on the final examination. He was interviewed by Goldschmidt, who said "Mr. Parr, you have not done well in my course; what do you intend to become?" Parr explained that he intended to become a marine biologist. Goldschmidt then said "Mr. Parr, if you will promise never, never to take another course in mineralogy, I will pass you," a bargain that Parr was glad to accept.

I had a very amicable relationship with Albert Parr during my time at the American Museum of Natural History; we enjoyed recalling Goldschmidt stories. A favorite one concerned a student's oral examination, and went as follows: Goldschmidt: "Tell me, Mr. X, why is amazonstone green?" (Amazonstone is a rare green variety of potassium feldspar.) Mr. X, trying to bluff it out, replied: "Well, sir, I did know but I have forgotten." Goldschmidt: "That's really too bad, Mr. X. Nobody else knows, and you have forgotten!" (It is now known that the green color results from defects in the structure produced by natural irradiation and induced by the presence of Pb^{2+} ions replacing K^{+}.)

In 1971 Professor Denis Shaw, editor of *Geochimica et Cosmochimica Acta* from 1971 to 1988, began gathering archival material on Goldschmidt, and corresponded with those people who had been associated with him. I am greatly indebted to Dr. Shaw for this material and the opportunity to quite from the correspondence, some of which has been incorporated in earlier chapters.

V. V. Shcherbina (1907–1978), Russian geochemist, recalls:[3]

> "I met V. M. Goldschmidt nearly every day from the middle of January until the middle of March 1933, when I was working in his institute under his direct supervision during my first scientific visit abroad in Germany (Berlin and Göttingen, September, 1932, to March 1933). Apart from a daily short oral report to Goldschmidt about my scientific work, I had the opportunity of meeting him and also questioning him about other problems in geochemistry. Frequently our discussions went beyond my topic of study (X–ray determination of the valency of V, Ti, and Fe in titanomagnetite and ilmenite) and Goldschmidt told me many interesting things. In particular, I recall the day when I was invited to his house for lunch—among the co-workers of his institute this was a common occurrence. A long scientific discussion ensued which was simultaneously my examination.
>
> Once a week, in the institute's auditorium, Goldschmidt gave a two–hour geochemical lecture which was full of facts and ideas. Not many students of the university attended these lectures, since at that time Germany did not recognize a need for this new type of geologist. All his scientific coworkers (which included 12–15 people) attended these lectures, along with students from abroad—M. Balkoni from Italy, H. Berman from the U.S.A., and myself from Leningrad."

Professor Ivar Oftedal, a friend and colleague of Goldschmidt for nearly thirty years, wrote:[4]

"The totally dominant feature of the person Goldschmidt was his nearly incredible mental faculties: his fund of knowledge, as well as his memory of details, was truly enormous, and everything was fully ordered in his brain and ready for use at any instant. This gift was invaluable not only to himself, but to anyone who asked for his advice and help in scientific matters, for he always gave his answers gladly and promptly. I have never heard him say "I have no time now, wait until tomorrow." During such conversations with Goldschmidt I felt like a modest pupil before the great master; however, he was quite as unassuming himself and there was no feeling of distance between us. I have never had a similar feeling towards anyone. There are many manifestations of his astonishing mastery. Asking him about some specified matter he would usually refer you to one or more papers to be found in volume so–and–so, and you could look it up straight away. When the latest issue of a journal arrived I could see him turning the pages slowly and continuously during some minutes, and after that he obviously had a good idea of the essential contents, which were then stored in his brain. Although Goldschmidt knew almost everything I think mathematics were somewhat alien to him. At least he was highly impressed by the mathematical abilities of Zachariasen, and even by my own casual application of elementary calculus in crystal structure work.

Goldschmidt was very kind–hearted and always willing to help where he could. It is important to remember this in view of certain facts which seem to contradict it. Combined with his kindness and friendliness was a high degree of suspiciousness and also a nearly explosive temper. I believe he had little knowledge of human nature and therefore felt uncertain towards new acquaintances. I have heard him say "I am glad for every person I do not get acquainted with." His temper did not show up very often, but sometimes he got into quite uncontrolled fury, expressing his rage and contempt in exaggerated terms which the "victim" could hardly take seriously. In most such cases I think Goldschmidt had misunderstood or at least exaggerated the situation and he often begged pardon afterwards. It was impossible to work with Goldschmidt for anyone who took his occasional scoldings seriously and got offended. The situation might certainly be very unpleasant indeed, but usually the sun was shining again the next day or sooner. But it is true that some of his "victims" did never get on friendly terms with him again. Goldschmidt himself was not very willing to forget about more serious quarrels, and therefore he had a number of "enemies," not least among well–known scientists. I do not usually know the reasons for his hatred towards certain persons, but sometimes he actually boasted about such situations. I have heard himself tell about several cases. During a discussion at a meeting of our Academy of Science he got so excited that the wound after a recent operation opened up ("I got so angry that I burst"). Also, at a faculty meeting in the university many years ago he got irritated over a statement by a woman professor: "Now we must finally take care to get rid of this damned petticoat government." He liked to tell us that the lady got so angry that she spat like a cat. This aspect of Goldschmidt's personality was well known among scientists and others, and it was often commented on, mostly with a smile. A distinguished Swedish professor used to ask, when meeting Norwegian geologists: "Now, has Goldschmidt got any new enemies?". It is true that his suspicious mind could easily destroy his contact with other people, but I believe the aversion was mostly on his own side—the "enemy" did not usually feel an equally intense hatred towards him. It was difficult for an "enemy" to restore good terms with Goldschmidt, but I have seen cases where this actually happened, especially in his later years. You will see that even in Goldschmidt's attitude towards his "enemies" there was an element of humour. In fact he was very fond of making jokes (not what you call practical jokes), sometimes highly grotesque, sometimes offensive, but mostly quite harmless or even cheap. For example: an excursion led us close to a rifle range, and suddenly we heard shooting. Then Goldschmidt said "If a bullet comes along you must step aside." To us Norwegians Goldschmidt's various remarks had an extra comic flavour; he knew Norwegian perfectly well, but he had a curious accent and used an almost literary syntax.

When assistants were carrying out laboratory work for his research programme he was often impatient like a child. He might stand outside the darkroom door during film development, shouting now and then: "How is it getting on? Have you soon finished?" But his undisguised delight at every successfully completed research detail was a real reward to any assistant or coworker.

My conversations with Goldschmidt scarcely touched on purely human, social, or political matters, so I know very little about his general attitude. Apart from his predominantly kind and benevolent nature I noticed his very strict obedience to law, his perfect honesty and reliability. He had a certain gift of foresight, which admittedly sometimes seemed too pessimistic to me. All the same he said, several months before World War II, that large-scale fission of uranium would certainly come before long, and that Hitler's troops would certainly invade Norway" (original English).

Mrs. L. W. Strock, fifty years after her stay in Norway, wrote:[5]

"I would like to include a few words on what a wonderful person Prof. Goldschmidt was to my husband and me. He befriended us in Göttingen even tho we had never met him before. In order to have my husband as a research co-worker, he paid for our living expenses the first six months until my husband was able to obtain a research grant for one year from U.S.A. He prolonged our stay for almost three years until Hitler made life so unbearable for the Prof. that he finally left Göttingen in Sept. 1935 and went to Oslo, Norway, making my husband promise that he would join him there as soon as possible. It wasn't until May 1937 that we came to Oslo, bringing along our 11–month old daughter. He gave us a royal welcome—treated us like we were close relatives. The plan was that we were to stay for three years, but we were advised to leave earlier and did so in Dec. 1938."

Dr. Leiv Gjessing (see Appendix C) gave a memorial address on Goldschmidt (VMG) to the Videnskaps–Akademi on January 27, 1988, the centenary of his birth, and has kindly provided the following translation.

"In 1939 I studied chemistry at the University of Oslo. When I was reading about the lanthanide contraction I asked a fellow student "What is that?" He answered "It is not part of the usual curriculum and therefore without interest. But if you absolutely wish to know more about it you should go and listen to the lectures of Professor Goldschmidt." So I did.

VMG was 51 years old in 1939 and gave lectures on crystal chemistry and crystal structure in such a way that I became fascinated both with him and his subjects.

One day at the end of his lecture VMG mentioned that if any student was interested in crystal structure analysis he was welcome at his institûte at the Geological Museum, where he could learn how to synthesize crystals and measure their structure by means of X–rays. Of the 20 students present I was the youngest, only 20 years old, with only one year of chemistry and physics for medical students and one year of pure chemistry. But I was the only one who, with great pleasure, accepted this exciting offer. The world—famous VMG was going to be my tutor.

The foreign students of VMG left Norway in the fall of 1939, due to the war, and I became his only Norwegian student, doing research at his institute from September 1939 to July 1940. Strangely enough Brian Mason came from New Zealand to VMG in January 1940, but he left at the beginning of April the same year. Unfortunately we never met. That was probably due to the fact that I studied mathematics and philosophy at the same time and did most of my laboratory and X–ray work in the afternoon and evening.

VMG instructed me how to make $CaNiSi_2O_6$ crystals and compare the structure with that of diopside, $CaMgSi_2O_6$. On the proposal of VMG the results were published in *Norsk Geologisk Tidsskrift* in 1940: "Contributions à l'étude des metasilicates."

VMG treated me very well, not only as his pupil but also as his friend. He gave me his classic work *Die Kontaktmetamorphose in Kristianiagebeit,* dedicated with the wish that I should not forget geology. He invited me to join him at the meetings of the Academy of Sciences, where he introduced me to the Nobel Prize winner George de Hevesy and other famous scientists. Later he was so kind as to engage me in the preparation of the third edition of his lectures in mineralogy, and in addition he hired me as his assistant during his lectures.

The work with crystal structures enhanced my interest in this field, and in July 1940 my friends Tollef Larsson and Harald Major joined me in a new project to exchange aluminum with indium, gallium, chromium, and iron in the mineral chrysoberyl, $BeAl_2O_4$. We formed a sort of triumvirate, under the guidance of VMG, resulting in the publication of the paper "Isomorphous substitution for Al^{3+} in the compound $BeAl_2O_4$" in the *Norsk Geologisk Tidsskrift* for 1942. I also passed the examination in mineralogy and crystallography.

VMG arranged many geological excursions in the Oslo region, studying the contact metamorphism and visiting iron, copper, and bismuth mines. We were also invited to parties in his home. On these occasions he was very lively, charming, and witty, telling interesting stories and funny jokes.

In the fall of 1942 VMG was arrested by the Germans, but he had a heart attack in the prison camp and was therefore admitted to a hospital. Later he escaped to Sweden, and from there he got to England by air. He returned to Norway in June 1946, but he was then weakened by his heart disease and the many difficulties during and after the war related to his return to Norway. One evening he became very ill with pulmonary edema due to heart failure. He was admitted to the National Hospital in Oslo in the middle of the night. I, as a medical student, happened to be on duty that night. I was very sorry to see my good friend and teacher in such a miserable condition. I was thankful to have this opportunity to help him, as he had helped me so much and so frequently over many years.

In 1946 I mentioned to VMG that I was going tramping in Iceland, Scotland, and England. He gave me the addresses to many of his good friends and asked me to visit them and bring greetings from him. That brought me in contact with many famous scientists. The Nobel Prize winner Max Born and his wife Hedi who lived at that time in Edinburgh, the geneticist Professor John Burdon Haldane, who brought me in contact with Professor S. L. Penrose, and Dame Kathleen Lonsdale, the crystallographer, in London. They were all very kind to me, and very much interested to hear news of VMG, about his health, and how he was adapting himself to Norway again.

When I went to Italy in 1947, VMG asked me to visit the Papal Observatory at Castel Gandolfo outside Rome. There I had the pleasure of meeting Father Junke and his coworkers in their spectrographic laboratory, famous for their work on the spectrum of the sun.

We who were his close pupils admired and respected him, both as a warm human being and as an eminent scientist. We well understood that his intellectual capacity was at an unusually high level. His classmate in high school, Dagfinn Dahl, characterized him in this way: "VMG learned everything quick as lightning, understood everything, and remembered everything."

Professor Victor Moritz Goldschmidt was a genius. In addition, he was an outstanding teacher and a good and faithful friend. We are very grateful for having been his pupils."

★ ★ ★ ★ ★

VICTOR MORITZ GOLDSCHMIDT (1903). Pencil and chalk drawing made in 1903 by Asta Elisa Jakobine Nörregaard (1853–1933), a Norwegian historical painter who lived in Oslo and studied under Knud-Larsen Bergslien. Provenance: Mineralogisk-Geologisk Museum, University of Oslo. Photograph courtesy J. A. Dons.

Victor Moritz Goldschmidt
(1888–1947)

Plate 1. Norwegian postage stamp issued in 1974 honoring V. M. Goldschmidt.

Plate 2. Heinrich Jacob Goldschmidt (1857–1937); father of V. M. Goldschmidt. He had succeeded Peter Waage in 1901, as Professor of Chemistry at the University of Oslo.

Plate 3. Photograph taken at a party in the home of Professor Heinrich J. Goldschmidt, father of V. M. Goldschmidt, in Oslo, February 1904. In attendance (left to right): Professor A. Nias; Professor J. H. Leonard Borgström (Lecturer of Mineralogy in 1904 at Kingston University; later Professor of Mineral Chemistry at Helsinki University, 1918–1947); Paul (?) (from Australia?); Dr Lincio; Dr P. F. Hubrecht; and Dr C. W. Dickson from Canada. Photo obtained from the archives of Professor P. Eskola courtesy of Professor Martti Lehtinen, Helsinki University.

Plate 4. Same location, individuals and acknowledgements as in Plate 3. The head of V. M. Goldschmidt is visible between Dr Lincio and Dr Hubrecht. At this time V. M. Goldschmidt was 16 years old. This is the earliest photograph located of V. M. Goldschmidt.

Plate 5. V. M. Goldschmidt (right) and his assistants, Mimi Johnson and Endre Berner, 1915. Photo courtesy J. A. Dons.

Plate 6. In 1920 Einstein spent several weeks in Oslo. On June 20 V. M. Goldschmidt arranged a geological excursion on the Oslofjord in the Geological Museum's motorboat; from left, Heinrich Goldschmidt, Einstein, Jacob Schetelig. As a result of this excursion Einstein took up sailing as a hobby, which he enjoyed for the rest of his life. This photograph and the following two were taken by H. Rosendahl (Curator, Geological Museum, Oslo).

Plate 7. V. M. Goldschmidt and Albert Einstein (in long black coat) at one of the islands in Oslofjord where they were looking at lower Paleozoic sediments. Goldschmidt is holding a specimen(s) which he is in the process of showing Einstein. (Photo taken at the same time and place as Plate 8).

Plate 8. V. M. Goldschmidt and Albert Einstein (long black coat) at one of the islands in Oslofjord, where they were looking at lower Paleozoic sediments; lunch table is shown at lower left.

Plate 9. V. M. Goldschmidt (right) with W. C. Brøgger (center) and J. Schetelig (discoverer of the first Sc mineral) on the porch of the Geological Museum in Oslo. The large specimen consists of inclusions of Precambrian gneiss in Permian dolerite. Photo taken in 1920.

Plate 10. V. M. Goldschmidt on field trip in the Stavanger area, 1922.

Plate 11. V. M. Goldschmidt and Professor F. Becke (Vienna) on field trip in the Stavanger area, June, 1922.

Plate 12. Portrait of V. M. Goldschmidt taken in 1923.

Plate 13. Artist's engraving of V. M. Goldschmidt (based on portrait shown in Plate 12) which appears on a Norwegian postage stamp issued in 1974 (see Plate 1).

Plate 14. Geologists from Norway and other countries at a seminar organized and led by V. M. Goldschmidt at the Mineralogical-Geological Museum, Oslo, in March 1923. The electric light bulb overhead was the source of the illumination for the microscopes by means of mirrors. The round table was designed by Goldschmidt and he would sit in the middle from which point he would assist the students. Identification of the individuals (from left to right) were supplied by J. A. Dons. *Standing:* Thorolf Vogt (Professor, Technical University, Trondheim; son of J. H. L. Vogt); Per Geijer (Professor, Technical University, Stockholm; later Director General of the Geological Survey of Sweden); Ivar Oftedal (later Professor of Mineralogy, University of Oslo); Olaf Anton Broch (geologist, Norwegian Geological Survey); Nils-Henrik Kolderup (later Professor, Bergen); H. Schneiderhöhn (in white; economic geologist, Germany; co-author with P. Ramdohr of *Lehrbuch der Erzmikroskopie* I, II 1931, 1934, Berlin); Jean Mette (representative of the optical industry); and V. M. Goldschmidt. *Seated:* Aarne Laitakari (Instructor of Geology, Technical University, Helsinki; later Director of Geological Survey of Finland, 1936–1960); Carl Bugge (Director, Norwegian Geological Survey); Steinar Foslie (geologist, Norwegian Geological Survey); Hans Henrik Hornemann (mining engineer); Olaf Andersen (co-worker with Goldschmidt and later with N. L. Bowen at the Carnegie Institution of Washington); and Heikki Vayrynen (assistant at Helsinki University; later Professor of Geology, Technical University, Helsinki). Several of the individuals (Vogt, Oftedal, Broch) will be seen in later photographs.

Plate 15. Group photo taken at the Geological Museum, Oslo, on 29 October 1928, the day Dr W. H. Zachariasen (later a professor at the University of Chicago) received his Ph.D. Photograph was taken by Lily Monsen (Geological Museum; secretary, photographer, librarian). Identifications of the individuals have been made mainly by J. A. Dons. 1. Johanne Nitter (drafting technician who later married Professor Sömme, geographer, University of Bergen); 2. Ivar Oftedal (curator, later Professor of Mineralogy); 3. Halvor Rosendahl (curator); 4. T. Ito (Professor of Crystallography, University of Tokyo); 5. Leif Störmer (later Professor of Paleontology); 6. Liv Barstad (drafting technician); 7. Pettersen (caretaker); 8. Ragni (Mossa) Durban-Hansen (drafting-photographic technician; granddaughter of W. C. Brøgger; later married Zachariasen in 1930; presently living in Santa Fe, New Mexico); 9. Ingeborg Strömme (secretary and librarian); 10. Alf Granli (technical curator); 11. Jahrl Andersen (technician); 12. Fesefeldt (spectroscopist, visitor from Germany); 13. Olaf Anton Broch (geologist, Norwegian Geological Survey); 14. Dorothy Wyckoff (guest worker, later Professor of Crystallography at Bryn Mawr College, Pennsylvania); 15. Miss Bredrup (secretary to V.M.G.); 16. Mimi Johnson Höst (assistant to V.M.G., specialist in microdocimastic preconcentration of noble metals); 17. K. Stenvik (X-ray technician); 18. Thorolf Vogt (Professor, Technical University, Trondheim); 19. Jacob Schetelig (Professor and Director of the Mineralogical-Geological Museum, Oslo); 20. W. H. Zachariasen; and 21. V. M. Goldschmidt.

Plate 16. V. M. Goldschmidt sitting in his office in the Mineralogisk-Geologisk Museum. The photograph was taken in the middle-late 1920s; certainly before 1929.

Plate 17. V. M. Goldschmidt's institute in Göttingen, 1929.

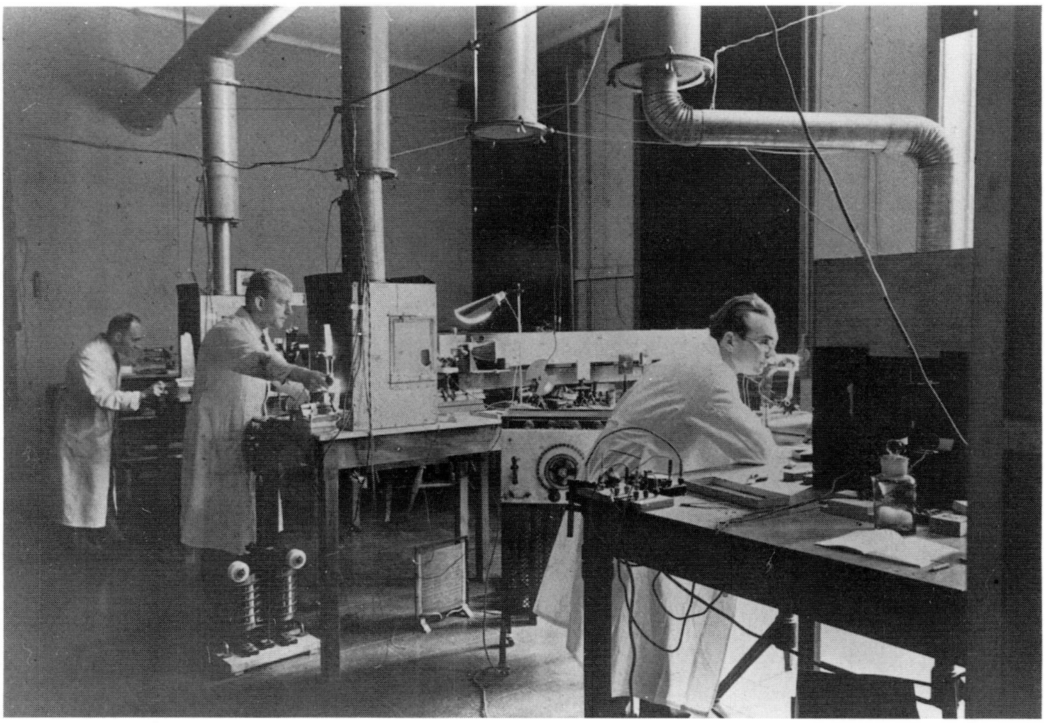

Plate 18. The optical spectrographs in the Göttingen institute, with (from left). L. Strock, H. Bauer and H. Hörmann.

Plate 19. Some of V. M. Goldschmidt's staff and associates outside the Göttingen institute, 1930. Back row (from left): C. Peters, F. Laves, K. Moeller, F. Machatschki; middle row (from left): Miss Grimel, Schwarz von Bergkampf (?), Mrs. Machatschki, H. Schumann, Miss Rakebrand (secretary); front row: W. Noll (?) and H. Schnaase. Identifications thanks to Professor Hans Wedepohl.

Plate 20. V. M. Goldschmidt and his father with Gulbrand Lunde, a friend (at the time) and co-author with V. M. Goldschmidt on several papers. Lunde, a Norwegian who studied in Germany as a University of Oslo scholar for many years up to 1924, later became a Nazi collaborator. During the war he became the counterpart of Goebbels in the Quisling government after the invasion of Norway. This photograph was taken in Göttingen sometime in the period 1929–1930.

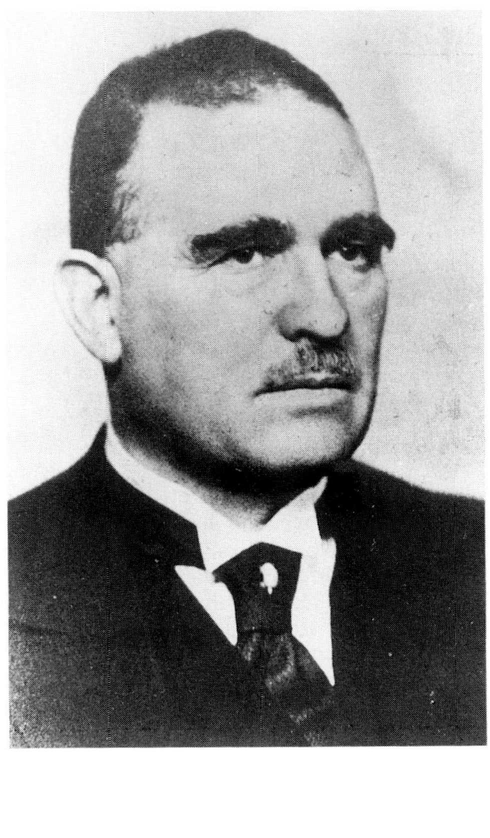

Plate 21. Studio photograph of V. M. Goldschmidt, 1932.

Plate 22. Group picture taken in front of the main entrance of the Mineralogisch-Petrographisches Institut, Göttingen, Lotzestrasse 16/18, in September 1931. Some of the individuals shown are from neighboring universities who had come for a course in X-ray crystallography. Identifications courtesy of Professor K. H. Wedepohl. 1. F. K. Drescher-Kaden (mineralogist; guest from Clausthal Mining School); 2. C. Hermann (crystallographer whose name is associated with "Hermann-Mauguin symbols", guest from Marburg University); 3. H. Ehrenberg (economic geologist); 4. V. M. Goldschmidt; 5. K. Dornberger-Schiff (crystallographer); 6. E. Reuning (economic geologist, guest from Giessen University); 7. H. Schnaase (coworker of V.M.G.); 8. H. Goedecke (coworker of V.M.G.); 9. S. Kyropoulos (solid material physicist from Greece); 10. K. Möller (coworker of V.M.G.); 11. Martin; 12. E. Broch (chemist, guest from Norway); 13. Th. Ernst (assistant; only student when Goldschmidt arrived at Göttingen in 1929); 14. H. O'Daniel (crystallographer); 15. G. E. R. Schulze (coworker of V.M.G.); 16. H. Hauptmann (visiting scientist); 17. R. Mannkopff (physicist who, with C. Peters (20), designed the spectrographs and initiated the sensitive methods for arc emission spectroscopy); 18. F. Buschendorf (mineralogist and economic geologist, guest from Clausthal Mining School); 19. F. Laves (crystallographer; later Professor at the University of Chicago and subsequently at ETH, Zürich); 20. C. Peters (chemist); 21. L. L. Quill (crystallographer, visiting scientist from U.S.A.); and 22. Novak (guest from Czechoslovakia).

Plate 23. V. M. Goldschmidt and W. Vernadsky in front of Goldschmidt's home (Wagnerstrasse 8, Göttingen) during the latter's visit in June 1932.

Plate 24. V. M. Goldschmidt and his father at the 92nd Conference of the *Gesellschaft Deutscher Naturforscher und Ärtze* in front of the Conference building, Mainz-Wiesbaden, 25–29 September 1932, where Goldschmidt gave a lecture on general crystal chemistry on 26 September.

Plate 25. V. M. Goldschmidt on field trip (1932/33) from the Göttingen Institute to the Habichtswald basalt area near Kassel, Germany. Identifications (by Professor K. H. Wedepohl) are, from left to right: Th. Ernst; F. Buschendorf; unidentified (with cane); unidentified (barely visible behind person with cane); Harry Berman (prominently in the center wearing a cap; Harvard University); unidentified (at rear); W. E. Trommsdorf (bending); V. M. Goldschmidt; and F. Laves.

Plate 26. V. M. Goldschmidt in Göttingen sometime in the period 1929–1935.

Plate 27. V. M. Goldschmidt in field clothes, Göttingen period (1929–1935).

Plate 28. V. M. Goldschmidt and R. Pohl (physicist) on an excursion in the Habichtswald, near Kassel, 1934. The notice reads: "It is forbidden to approach the ruins because of danger to life".

Plate 29. V. M. Goldschmidt, Mrs L. W. Strock and Dr E. Minami (visiting scientist from the University of Tokyo) at the Göttingen railroad station, in July 1935 on the occasion of Dr Minami's departure for Japan after a stay of about two years. Dr Minami, in 1935, was the first person to publish rare-earth element concentrations in common rocks, specifically shale composites from Europe and Japan; this work was done under the direction of Goldschmidt.

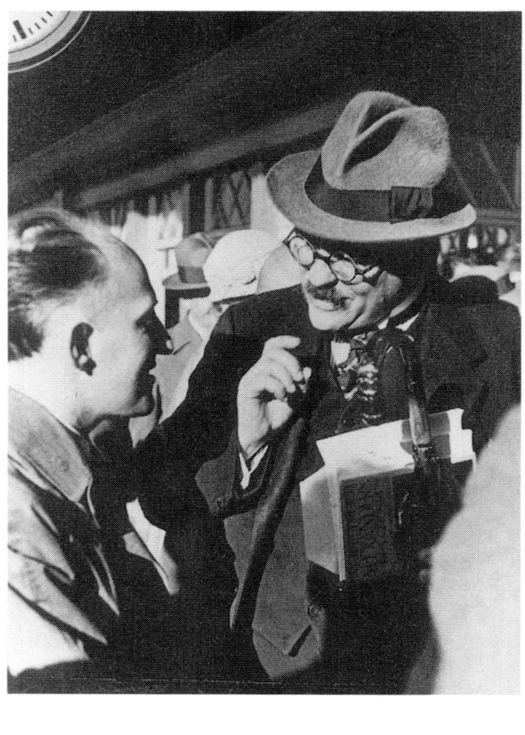

Plate 30. V. M. Goldschmidt saying goodbye to Dr L. W. Strock at the Göttingen railroad station at 5:00 PM on 6 September 1935 as Goldschmidt departed for Norway because of the deteriorating political situation and rising anti-Semitism in Germany; Goldschmidt is also giving Strock last minute instructions about Strock joining him in Oslo. Dr Strock left Germany to return to the United States (via London where he worked for several months) in November of that year (1935). He rejoined Goldschmidt in Oslo in May, 1937 and remained until December 1938 (see Plate 35).

Plate 31. The Holmenkollen house where V. M. Goldschmidt (and the Barth family) lived from 1936 to 1940.

Plate 32. V. M. Goldschmidt and Assar Hadding (Professor of Geology, Lund University) on a field trip in southern Norway, June 1936.

Plate 33. V. M. Goldschmidt's father walking with "Highland Bob", winter 1936–37.

Plate 34. V. M. Goldschmidt on field trip to the Röros mining area, central Norway, in 1937.

Plate 35. Farewell tea for Dr Lester W. Strock and family taken at the Geological Museum (in the same room as Plate 15), Oslo, in December 1938. Identification of the individuals have been made mainly by J. A. Dons. 1. Ivar Oftedal (mineralogist, curator, later Professor); 2. Kaare Münster-Ström (limnologist, later Professor); 3. Åsta Ingebrigtsen (wife of caretaker); 4. T. Myhren; 5. Borgny Bay (drafting technician); 6. K. Stenvik (X-ray technician); 7. Unn Hofseth (drafting technician); 8. Liv Ingebrigtsen (daughter of caretaker); 9. Ole Larsen; 10. Anatol Heintz (paleontologist, later Professor and Director of the Paleontological Museum); 11. Eisland; 12. Jarhl Andersen (technician); 13. H. C. Ingebrigtsen (caretaker); 14. Bitte Falck (secretary); 15. Aslak Kvalheim (Director of the Raw Material Laboratory following V.M.G.); 16. Mrs L. W. (Mary) Strock; 17. Miss Winifred Strock; 18. Lester W. Strock; 19. Alf Granli (technical curator); 20. K. Kristoffersen (curator); 21. V. M. Goldschmidt; and 22. Leif Störmer (paleontologist, later Professor).

Plate 36. V. M. Goldschmidt and his dog at a dog show (competition) in Oslo. The photograph was probably taken in about 1938; at least this was the dog he had in 1938.

Plate 37. V. M. Goldschmidt on student field trip to the best contact metamorphism locality (where he did his original research for his Ph.D. thesis), Årvollåsen, Norway, in 1939. From left: Leif Gjessing, G. Henningsmoen, Alf Granli and Tore Gjelsvik. Photo by J. A. Dons.

Plate 38. Group photo taken at the Geological Museum, Oslo, on 25 April 1939 when former group member Dr W. H. Zachariasen (see Plate 15), then of the University of Chicago, returned for a visit. Identifications of the individuals have been made mainly by J. A. Dons. 1. Mrs Arneberg (secretary); 2. Leif Störmer (paleontologist, later Professor); 3. Kaare Münster-Ström (limnologist, later Professor); 4. K. Kristoffersen (curator); 5. W. H. Zachariasen; 6. Unn Hofseth (drafting technician); 7. Mrs Zachariasen; 8. Anatol Heintz (paleontologist, later Professor and Director of the Paleontological Museum); 9. V. M. Goldschmidt; 10. Nina Heintz [secretary, sister of (8)]; 11. K. Stenvik (X-ray technician); 12. Olaf Holtedahl (geologist and later Professor; author of *Geology of Norway*); 13. Ivar Oftedal (mineralogist, curator and later Professor); 14. Aslak Kvalheim (Director of the Raw Material Laboratory after V.M.G.); 15. Halvor Rosendahl (curator); 16. Alf Granli (technical curator); 17. Bitte Falck (secretary); and 18. T. Myhren.

Plate 39. V. M. Goldschmidt examining contact metamorphism. Årvollåsen, Norway, in 1939. From left: G. Henningsmoen and Kare Breivik. Photo by J. A. Dons.

Plate 40. Photograph of V. M. Goldschmidt taken in Stockholm in January 1943.

Plate 41. V. M. Goldschmidt walking in the procession to receive an honorary LL. D. degree from the University of Aberdeen, June 29, 1944. The other honorees (from left) are Professor Alexander Findlay, Colonel James Dawson, the Right. Rev. Herbert Hall (Bishop of Aberdeen and Orkney), and the Rev. John Anderson.

Plate 42. A passport photograph of V. M. Goldschmidt taken (by Lafayette) in 1944 while he was living in England. From "Obituary Notices of Fellows of The Royal Society", Vol. VI, pp. 51–66, 1948–1949.

Plate 43. The urn (36 cm in height), made of beautiful green olivine rock, in which V. M. Goldschmidt's ashes were placed.

Notes

1/Prologue

1. In 1939 £1 (NZ) was approximately equivalent to \$4 (US) or 16 Norwegian kroner.
2. Pabst, Adolf (1899–1991), American mineralogist, B.A., University of Illinois, 1925; Ph.D. (geology), University of California, 1928. American–Scandinavian Foundation Fellow, University of Oslo, 1928–29. Professor of Mineralogy, University of California (Berkeley), 1929–67. President, Mineralogical Society of America, 1951, Roebling Medallist, 1965.
3. Barth, Thomas Fredrik Weiby (1899–1971), was the son of a highway engineer and grew up in Trondheim, where he began his university studies at the Norwegian Technical University in 1918. He made his first acquaintance with Goldschmidt in the winter of 1919–20, when he worked at the Geological Museum assisting the visiting Finnish scientist Pentti Eskola in the chemical analysis of rocks and minerals. In the autumn of 1919 he enrolled at the University of Oslo, from which he received his doctor's degree in September 1927. From the autumn of 1924 to the spring of 1927 he worked as a scientific assistant to Goldschmidt, for whom he made over 1000 X–ray diffraction photographs and was author or co–author on 19 publications. In a recommendation written in March 1927 Goldschmidt wrote: "His remarkable scientific competence, his skill and diligence are clearly shown by his extensive series of excellent publications. I greatly appreciate his personal qualities, as an extremely reliable, conscientious and gifted co-worker." After graduation in 1927 he worked for some time in Germany, before going to the U.S. in the spring of 1929, first to Harvard University, and from the autumn of 1929 at the Geophysical Laboratory in Washington, D.C. He later remarked that his years at the Geophysical Laboratory were the happiest

in his life. In February 1936 he returned to Norway as docent in crystallography, mineralogy, and petrology at the University of Oslo, and was promoted to Professor in July 1937. He returned to the Geophysical Laboratory in September 1939, coming back to Oslo in April 1940 immediately before the German invasion. In June 1946 he accepted an appointment as Professor of Petrology at the University of Chicago, but returned to Norway in March 1949 when he was appointed Director of the Geological Museum, a position which he held until retirement in 1966. He was active in international affairs, being president of the International Union of Geological Sciences from 1964 to 1968, and he travelled widely throughout the world, frequently as a guest lecturer. He received many honors, among them the Roebling Medal, the premier award of the Mineralogical Society of America, in 1961.

Although his given name was Thomas, he always used the abbreviated form, but insisted that it be followed by a period when printed, a usage followed in this book.

4. The name on the facade is "Geologisk Museum"; the building houses the Mineralogisk–Geologisk Museum and the Paleontologisk Museum. "Geological Museum" will be used throughout this book, mostly referring to the first–mentioned. The museum is part of the University of Oslo.
5. Marie Brendingen (the name is also spelled Brendengen, Braendengen, and Brenningen; in his correspondence Goldschmidt spelled it Brendingen, and that usage is followed here) was born in 1891 and died in 1980. She joined the Goldschmidt household as a housekeeper in 1927 and ministered faithfully to him until his death in 1947. She became more than a housekeeper; she was devoted to him and provided a comfortable home where he was able to concentrate on his scientific and industrial work. He bequeathed a large part of his estate to her.
6. Lunde, Gulbrand Oscar Johan (1901–1942) was born in Bergen. He studied at the University of Freiburg in Breisgau (Germany) and received his doctor's degree in chemistry in 1924. From 1924 to 1928 he was employed by the Raw Materials Laboratory as scientific assistant to Goldschmidt, during which time he carried out extensive crystal chemical researches, resulting in some 20 publications as author or co–author. He did pioneer work on the geochemistry of iodine. From 1929 to 1940 he was director of the Norwegian Cannery Industry's laboratory in Stavanger, where he made a series of scientific investigations of importance to that industry. After Quisling founded the Nasjonal Samling (the Norwegian Nazi party) in 1934, Lunde became an enthusiastic member, and belonged to Quisling's inner circle. Quisling appointed him to his abortive "government" in April 1940, and in September

1940 he was named by Reichskommissar Terboven to the "kommissarisk statsråd" and chief for the Department of Information and Propaganda. When the Quisling government was appointed in February 1942 he became Minister of Propaganda, the Norwegian Goebbels. Lunde and his wife died in a drowning accident on October 25, 1942.

Goldschmidt greatly appreciated Lunde's scientific talents, and supported his election as a member of the Videnskaps–Akademi in 1938. After Lunde revealed himself as a traitor in April 1940 Goldschmidt neither met nor corresponded with him again.

7. His silicosis experiments were aimed at substituting crushed olivine for quartz as a molding sand in foundries. The use of quartz resulted in serious lung damage to the workers.

2/Early Years: 1888–1906

1. Meyer, Victor (1848–1897), German chemist, Ph.D., University of Heidelberg, 1867. Professor of Chemistry, Eidgenossische Polytechnikum Zürich, 1872–85, where he was a colleague of Goldschmidt's father. Professor of Chemistry, University of Göttingen, 1885–89; University of Heidelberg, 1889–97.
2. van't Hoff, Jacobus Henricus (1852–1911), Dutch chemist, Ph.D. Utrecht, 1874. Chairman, Chemistry Department, University of Amsterdam, 1878–96; University of Berlin, 1896–1911. In Berlin he applied the Phase Rule to the elucidation of the mineralogy and paragenesis of salt deposits.
3. Rosbaud, Paul Wenzel Matteus (1896–1963) was born in Graz, Austria. In March 1915 he enlisted in the Austrian army and was sent to the Italian front. On November 3, 1918, he surrended to a British unit; his experiences as a prisoner of war made him a life–long Anglophile. In 1920 he entered the Technische Hochschule at Darmstadt in Germany to study chemistry, where he met his wife–to–be, Hilde Frank. He took his doctorate at the Technische Hochschule in Berlin. In January 1928 he became a scientific advisor to the journal *Metallwirtschaft*, and in this capacity met many of the leading scientists of the day, including Goldschmidt. In 1930 he joined the publishing house of Springer Verlag as scientific editor. In April 1938 he moved his wife and daughter Angela to England. In the late 1930's he established contact with the British Secret Intelligence Service (SIS), and during the war he was a British agent (code name Griffin) providing valuable technical data on German military secrets. He came under suspicion and in July 1944 was to be drafted into Organization Todt (slave labor), but was saved by the intercession of F. K.

Drescher–Kaden, Goldschmidt's successor at Göttingen. In recognition of his services he was invited to settle in England in November 1945, and he lived there till his death. He was instrumental in founding the international journal *Geochimica et Cosmochimica Acta* in 1950, and in the posthumous publication of Goldschmidt's *Geochemistry* in 1954. A book–length biography *The Griffin,* by Arnold Kramish, was published in 1986 (Houghton Mifflin Co., New York).

4. Brøgger, Waldemar Christopher (1851–1940) was born in Oslo and educated at the Cathedral School and the University of Oslo. He began his scientific career as a zoologist, his first publication in 1873 being a description of the molluscan fauna of the Oslofjord. He soon turned to geology, and he had a distinguished and versatile record in mineralogy, petrology, paleontology, and stratigraphy. In 1881 he was appointed the first professor of geology and mineralogy in Stockholm Högskola [now University of Stockholm]; during this period he described some fourteen new minerals and published a large monograph on the geology and mineralogy of the pegmatites of southern Norway. In 1890 he returned to Oslo as professor of mineralogy and geology in the university, a position he occupied until his retirement in 1916. Besides his scientific work, he was active in public affairs, as rector of the university (1907–1911), a member of parliament (1907–1909), and president of the Videnskaps-Akademi for some twenty years (1915–1935). He continued his researches after his retirement, publishing extensively until the mid 1930's. As teacher, colleague, and friend he was a major influence on the life of V. M. Goldschmidt.
5. Hiortdahl, Thorstein Hallager (1839–1925), Norwegian chemist and crystallographer, Professor of Chemistry, University of Oslo, 1872–1918. He studied relationships between chemical composition and crystal form, and made many difficult mineral analyses.
6. Birkeland, Kristian Olaf (1867–1917), Norwegian physicist, Professor of Physics, University of Oslo, 1898–1917. Developed with Sam Eyde a process for the oxidation of atmospheric nitrogen in an electric arc for the production of nitric acid and nitrates (the Birkeland–Eyde process). Commercial use of this process was initiated in Norway in 1907 by the firm Norsk Hydro.
7. Hevesy, George de (1885–1966) was born to a wealthy and distinguished Hungarian family. He received his early education in Budapest, but in April 1905 he entered the University of Freiburg, receiving his Ph.D. in physical chemistry in 1908. Goldschmidt spent several weeks in the Chemistry Department at Freiburg in the summer of 1906 and thus began a lifelong friendship. In 1911 Hevesy went to the University of Manchester to do research work in radioactivity in Rutherford's department. Here he met many

research workers, and was particularly impressed by Niels Bohr and H. J. G. Moseley, both of whom greatly influenced his future career. On the outbreak of World War I he returned to Vienna and joined the army, from which he was assigned to work in the armaments industry. After the war he accepted a position in the University of Budapest, but political difficulties persuaded him to accept Bohr's offer of a position in his institute in Copenhagen, from April 1, 1920. He then became greatly interested in geochemistry, applying X-ray spectrography to the analysis of rocks and minerals. This led to the discovery in 1922, with Dirk Coster, of element 72, which they named hafnium. He was appointed Professor of Chemistry at the University of Freiburg in April, 1926, but after the Nazis came to power in 1933 he resigned and returned to Bohr's institute in Copenhagen.

He next developed the use of isotopes as biological tracers, firstly with deuterium and later with a radioactive phosphorus isotope; the significance of this work was recognized by the award of the Nobel Prize for Chemistry in 1943. When the persecution of the Jews began in Denmark in 1943 he escaped to Sweden and accepted a research position in the University of Stockholm, where he remained until his retirement in 1961. He remained in close contact with Goldschmidt throughout the latter's life. Hevesy was a keen skier and took many vacations in the Norwegian mountains, usually staying for a few days with Goldschmidt in Oslo, and Goldschmidt frequently visited Hevesy in Copenhagen. There is extensive correspondence between them in the Goldschmidt Archives.

3/The Petrographic Years: 1907–1921

1. See Appendix F for the geological time scale.
2. Eskola, Pentti (1883–1964) was the son of a landowner in southwest Finland and received his doctoral degree in geology from the University of Helsinki in 1914. His doctoral thesis "On the petrology of the Orijärvi region in south-western Finland" applied Goldschmidt's principles of equilibrium in metamorphic rocks to an area of complex regional metamorphism. In 1919–20 he spent almost a year in Goldschmidt's institute in Oslo and developed his concept of the mineral facies of rocks. "The term facies to designate a group of rocks characterized by a definite set of minerals which, under conditions obtaining during their formation, were in perfect equilibrium with each other" (*Norsk Geologisk Tidsskrift,* vol. 6, p. 145, 1920). In his acknowledgements he wrote "Professor Goldschmidt has had a very effective share

in my work. During frequent colloquies with him most of the problems here treated were discussed, and many of the ideas developed have grown from his suggestions." The facies concept has been a guiding principle in elucidating the mineralogy and petrography of metamorphic rocks. Eskola was appointed professor of geology and mineralogy at the University of Helsinki in 1924, a position he retained until his retirement in 1953. Goldschmidt and he were fast friends, and corresponded frequently; Eskola sent many of his students to study and do research in Goldschmidt's institute.

3. Becke, Friedrich (1855–1931), Austrian mineralogist and petrologist, Ph.D., University of Vienna, 1880. Professor, University of Czernowitz, 1882–90; University of Prague, 1890–98; University of Vienna, 1898–1927. After Goldschmidt studied with him in the winter of 1908–09 they were friends for the rest of Becke's life. He is remembered by the widely-used Becke method for determination of the refractive indices of small grains by immersion in liquids of known refractive index.
4. Groth, Paul Heinrich (1843–1927) was educated at the Mining Academy at Freiberg in Saxony and at the University of Berlin. From 1872 to 1883 he was Professor of Mineralogy at the University of Strasbourg in Alsace, and then at the University of Munich, a position he held until his retirement in 1924. Both in Strasbourg and in Munich his institute was for many years the center of training for crystallographers and mineralogists from all parts of the world. He was an inspiring teacher, and he did more than anyone else in stabilizing crystallographic nomenclature and methods. He first met Goldschmidt on a mineral collecting trip in Norway in 1910 and they became fast friends; Groth proposed Goldschmidt as his successor at the University of Munich, which failed because of anti-Jewish feeling among the faculty.
5. Johnson, Mimi (1890–1980) graduated in mineralogy from the university in Oslo in 1912. She was appointed assistant in the museum in 1913 and was promoted to curator in 1917. She assisted Goldschmidt during the summers of 1913–17 in his field work in the Norwegian mountains. She married in 1919, becoming Mrs. Johnson-Høst. From 1924 to 1931 she was a research associate in Goldschmidt's laboratory, during which time she and Gulbrand Lunde developed and applied microchemical methods for the determination of trace amounts of gold, silver, and the platinum metals in rocks and minerals. In 1930 she spent four months in Göttingen teaching a course in petrographic microscopy. She then became a medical student, graduating in 1940 and practising for many years as a physician in a small town near Oslo.

6. Berner, Endre Qvie (1893–1983), field and laboratory assistant for Goldschmidt, 1913–17, later Professor of Chemistry, University of Oslo, 1934–62.
7. Amundsen, L., *Videnskapsakademi Historie,* vol. 2, p. 27. Oslo, 1960.
8. *J. Washington Acad. Sci.,* vol. 51, p. 69–76, 1961; translated from Norwegian by G. Kullerud.
9. In his obituary of Goldschmidt, C. E. Tilley commented: "(Goldschmidt) was unaware, it is clear, of the much earlier zonal work of Barrow in the southeast Highlands of Scotland." (*Obituary Notices of Fellows of the Royal Society,* vol. VI, 1948–49, p. 51). However, Barrow's work had been generally overlooked, perhaps because of the rather non-informative title of his publication—"On an intrusion of muscovite–biotite gneiss in the south–east Highlands of Scotland" (*Quart. J. Geol. Soc. London,* vol. 49, p. 330–358, 1893).
10. The Geological Museum in Oslo has a large map of the Stavanger–Trondheim area prepared by Goldschmidt, closely spaced with colored pins denoting rock types at specific localities. He used it in his lectures and had a special box made to transport it.

4/The Raw Materials Laboratory

1. Prior to the founding of the Raw Materials Laboratory, Goldschmidt had already developed some industrial contacts, as consultant to the Norwegian companies A/S Electroverk and A/S Titan. In 1916 his consultant fees equalled his professor's salary (7000 kr/year). When he was appointed Director of the Raw Materials Laboratory he gave up his consultancies, and the Commerce Department paid him an honorarium of 6000 kr/year. The amount of this honorarium remained unchanged until his death in 1947.

   The Raw Materials Laboratory produced a series of publications, most of which were issued as bulletins of the Geological Survey of Norway.
2. *Norsk Vidensk. Selsk. Skrifter. I. Math.-Naturv. klasse,* No. 9, 1921.
3. Laue, Max von (1879–1960), German physicist, who in 1912 discovered the diffraction of X–rays by crystals, for which he received the Nobel Prize in Physics in 1914. Professor of Physics, University of Zürich, 1912–14; Frankfurt am Main, 1914–19; Berlin, 1919–43; Göttingen, 1946–51. Director of the Fritz Haber Institute for Physical Chemistry, 1951–58.
4. Bragg, William Henry (1862–1942), British physicist, joint winner with his son, W. L., of the Nobel Prize in Physics in 1915 for the

determination of crystal structures by X–ray diffraction. Professor of Physics, University of Adelaide (Australia), 1886–1908; Leeds, 1908–15; London, 1915–25; Director of the Royal Institution (London), 1925–42.

5. Bragg, William Lawrence (1890–1971), British physicist, son of W. H. In 1919 he succeeded Rutherford as Professor of Physics, University of Manchester, where he built up a school distinguished for crystal structure determinations. Professor of Physics and Director of the Cavendish Laboratory, Cambridge University, 1937–53 (again in succession to Rutherford). Director, Royal Institution (London), 1953–66.
6. Bernal, John Desmond (1901–1971), British physicist and X-ray crystallographer, Professor of Physics (1938–63) and Crystallography (1963–68), University of London.
7. *J. Chem. Soc.* (London), p. 2108, 1949.
8. Oftedal, Ivar Werner (1894–1976) was born in Larvik, Norway. He began his studies at the University of Oslo in 1913. In 1918 he began his long association with the Geological Museum as a student assistant to Goldschmidt in mineralogy, eventually rising to the position of senior curator (and acting director during Goldschmidt's time in England and briefly after his death). During the 1920's he worked closely with Goldschmidt on crystal chemical investigations, but later developed his own interests in mineralogy and geochemistry. After Goldschmidt's death in 1947 he succeeded him as Professor of Mineralogy, until his retirement in 1964.
9. Thomassen, Lars (1896–1972) was born in Oslo and graduated in chemical engineering from the Norwegian Institute of Technology in Trondheim in 1919. From 1919 to 1924, and again from 1927 to 1929, he was Goldschmidt's principal research associate. He was largely responsible for the construction and operation of the X–ray spectrographs which played such a large part in the Oslo researches, and in 1926 he published a comprehensive account of this technique for the detection and measurement of concentrations of the chemical elements down to a level of 0.1%. In December 1924 he went to the California Institute of Technology in Pasadena on a fellowship from the International Education Board, to work with Professor R. A. Millikan. He presented a thesis and was awarded a Ph.D. before returning to Norway in September, 1927. In 1929 he was appointed professor in the Department of Chemical Engineering of the University of Michigan, specializing in metallurgy, where he remained until retiring in 1966.
10. Stenvik, Kristoffer (1892–1959) was born at Ytterøy in the Trondheimsfjord. At the age of 16 he began work as a mechanic and worked for industrial and mining companies until June 1921,

when he accepted an appointment as technical assistant in the Raw Materials Laboratory. He studied mineralogy, petrology, and chemistry at the university, and became Goldschmidt's right–hand man. They complemented each other in a remarkable manner—Goldschmidt with his theoretical knowledge and fertile ideas, and Stenvik with his technical expertise. They worked together closely throughout Goldschmidt's life, especially in the development of Norwegian olivine rock as a refractory, beginning around 1925. They took out several patents together. In October 1947 the company A/S Olivin was formed to exploit the olivine deposits in Norway, and Stenvik managed the operations for almost ten years, before returning to the Raw Materials Laboratory in 1957.

11. A more detailed Periodic Table is given in Fig. 7.
12. Hadding, Assar Robert (1886–1962), Swedish geologist and mineralogist, graduated from the University of Lund 1913 and appointed docent, advanced to professor 1934; Rector of the University of Lund, 1947–51. Hadding was a many–sided researcher who worked over practically the whole field of geological science. His pioneer work on the application of X–ray spectrography and X–ray diffraction to mineralogical and geochemical problems was an outstanding contribution, and was highly esteemed by Goldschmidt; they were good friends throughout his life.

5/Intermezzo: The Search For Element 72

1. This chapter is partly based on the article "The era of cosmochemistry and geochemistry, 1922–1935," by Gustaf Arrhenius and Hilde Levi, published in G. Marx (editor): *George de Hevesy, 1885–1966.* Akademiai Kiado, Budapest, 1988, pp. 11–36.
2. *Chemical Analysis by X-rays and its Applications,* p. 182. McGraw–Hill Book Co., New York, 1932.
3. Letter dated February 7, 1925, Translated from Norwegian by Professor Gustaf Arrhenius, University of California, San Diego.
4. The yearly demand for hafnium in the U.S. now exceeds 50 tons. It is used for control rods in atomic reactors, and to a minor extent as a component of special alloys.

6/Crystal Chemistry and Geochemistry in Oslo: 1922–1929

1. *Chemical Products,* p. 2, March–April 1944.
2. Zachariasen, Fredrik William Houlder (1906–1979) was born in Langesund in southern Norway, the center of one of the world's

great mineral localities. In 1923 he entered the University of Oslo and joined the team of young researchers in the Mineralogical Institute. He soon showed his remarkable talent in the X–ray analysis of crystal structures; he was a co–author on *Verteilungsgesetze* VI and VII, and in 1928 received his doctor's degree for a 168–page monograph *Untersuchungen über die Kristallstruktur von Sesquioxyden und Verbindungen ABO_3* (Investigations of the crystal structure of sesquioxides and compounds with the formula ABO_3.) After graduation he worked for some time with W. L. Bragg at the University of Manchester before joining the faculty of the Physics Department at the University of Chicago, where he remained until his retirement in 1974. Among his many contributions was his work for the Manhattan project on the crystal structure and crystal chemistry of the actinide elements—thorium, protactinium, uranium, neptunium, plutonium, americum, curium. Goldschmidt considered him one of his most gifted students and collaborators, and tried unsuccessfully to get him a research professorship in Norway in 1936.

3. Ulrich, Frantisek (1899–1941) was born at Boharyne in eastern Bohemia and was educated at the Charles University in Prague, receiving his doctor's degree in mineralogy. In 1923 he received a fellowship from the Rockefeller International Education Board and joined Goldschmidt's research group in Oslo. He was a co–author in Part IV of the *Geochemische Verteilungsgesetze* on the crystal structure of the rare earth oxides, he determined the structure of Al_2O_3, and collaborated with Zachariasen on the structures of greenockite (CdS) and wurtzite (ZnS). He returned to Prague in 1925 and organized an X–ray laboratory in the Mineralogical Institute of the Charles University; he was appointed Professor of Experimental Mineralogy in 1934. He was murdered by the Gestapo in 1941.
4. Welsbach, Auer von (1858–1929), Austrian chemist, was a foremost investigator of the chemistry of the lanthanides, and the inventor of the lighter "flints" which are an alloy of the lanthanide elements, and of the incandescent lamp mantle made of lanthanide and thorium oxides. Contemporaneously with Georges Urbain, he discovered element 71 which he named cassiopeium, but Urbain was awarded priority with his name lutetium.
5. Thus the cell edge of NaF, 4.62Å, is the sum of the diameters of a sodium and a fluoride ion. From Wasatjerna's figure of 1.33Å for the radius of the fluoride ion, it follows that the radius of the sodium ion is 0.98Å.
6. Today one can hardly visualize the experimental difficulties with X–ray investigations in the 1920's. X–ray tubes were temperamental and required constant attention, obtaining a satisfactory vacuum could take hours of pumping, films were slow so exposure times were long, and computers were non–existent (even in 1942,

when I was doing X–ray crystallographic research, all calculations were made on a hand–operated mechanical calculator).

7. *Chemical Products,* p. 4, March–April 1944.
8. *Norsk geol. Tidsskr.,* vol. 9, p. 258, 1928.
9. Machatschki, Felix Karl Ludwig (1895–1970) was born in Arnfels, Styria, Austria and educated in Graz, receiving his doctor's degree in mineralogy and petrology. From 1927 to 1930 he worked closely with Goldschmidt, except for a period in Manchester in 1928–29. During that time he published a landmark paper in which he enunciated the modern theory of the crystal structure of the silicates: "Zur Frage der Struktur und Konstitution der Feldspate (Gleichzeitig vorlaüfige Mitteilungen über die Prinzipien des Baues der Silikate," in *Centralblatte der Mineralogie,* Abteilung A, p. 97–104, 1928). As he acknowledges, this was based on the data on ionic radii and isomorphism in Goldschmidt's *Geochemische Verteilungsgesetze der Elemente.* In 1929–30 he was guest lecturer at Goldschmidt's institute in Göttingen. In 1930 he was appointed Professor of Mineralogy at the University of Tübingen, followed by similar positions at the University of Munich (1941–44) and Vienna from 1944. His great contributions to the crystal chemistry of minerals were recognized by numerous honors, among them the Roebling Medal, the highest award of the Mineralogical Society of America.
10. Wyckoff, Dorothy (1900–1982) was born in Massachusetts, the daughter of a Congregational minister. She entered Bryn Mawr College in 1917 and received the A.B. degree in 1921, majoring in Greek and Latin. After a period of school teaching she returned to Bryn Mawr in 1925 to study geology and related sciences. She received her M.A. in geology in 1928 and was awarded an American–Scandinavian Foundation Fellowship. She took up the fellowship at the University of Oslo in September 1928 and under the guidance of V. M. Goldschmidt made a field and laboratory study of the metamorphic rocks of the Mt. Gausta region in the province of Telemark, southern Norway. For her thesis on this work she received her doctorate from Bryn Mawr in 1932. In September 1930 she was appointed an assistant in the Geology Department at Bryn Mawr and was successively promoted, eventually to full professor in 1945. During World War II she joined the Military Geology Unit of the U.S. Geological Survey in Washington, D.C., where her talents as geologist, photogrammetrist, and artist enabled her to produce terrain diagrams which were used to plan assault operations and were famous for their accuracy and clarity. She retired from Bryn Mawr College in 1966.
11. The nomination was repeated in 1930, 1931, 1932, 1933, 1934, and 1936, without success.
12. Eskola Archives, University of Helsinki.

7/Geochemistry in Göttingen: 1929–1935

1. W. A. Benjamin Inc., New York, 1965 (p. 363).
2. *Teknisk Ukeblad,* p. 124–125, 1936. The building is now part of the Physics Department, but a plaque on the outside wall records Goldschmidt's tenure. The Geochemical Institute moved to a new building on the outskirts of Göttingen in 1971, the address being appropriately Goldschmidtstrasse 1. It houses a collection of some 2500 documented rock and mineral samples analysed by Goldschmidt and his co-workers, together with the spectrographic plates.
3. Heide, Friedrich (1891–1973) was born in Dresden and was a student at the universities of Munich and Jena during the years 1911–1920, with an interval of military service during World War I. After graduation he was an assistant at the University of Jena until 1923, after which he moved to the University of Göttingen, first as assistant and later as docent in the Mineralogical Institute. After the retirement of Professor Mügge in 1928, he was in charge of the Mineralogical Institute until Goldschmidt took up the directorship in late 1929. During this time he supervised the construction and equipping of the new institute to Goldschmidt's specifications, and developed his interest in geochemical research. In 1930 he returned to Jena as Professor of Mineralogy and Director of the Mineralogical Institute, a position he occupied until his retirement in 1963. He is particularly noted for his research on meteorites, probably inspired by his association with Goldschmidt; his book *Kleine Meteoritenkunde,* first published in 1934, has gone through several editions and was translated into English by Edward Anders (a new version is in preparation by F. Wlotska and R. S. Clarke, Jr.).
4. Ernst, Theodor (1904–1983) was a student at Göttingen when the negotiations for Goldschmidt to accept the professorship began in 1927. In March 1929 he was appointed a teaching assistant in the institute, and worked closely with Goldschmidt in the lecture and laboratory instruction. He received his doctor's degree in petrology in 1936, and in 1937 was appointed docent in mineralogy and petrology. He remained at Göttingen until 1949 (except during World War II), when he was appointed Professor of Mineralogy and Geology at the University of Erlangen. He retired in 1972.
5. Gleditsch, Ellen (1879–1968), Norwegian chemist, was educated in Oslo and worked with Madame Curie in Paris from 1907 to 1912 and with Professor B. B. Boltwood at Yale University 1913–14, when she received an honorary D.Sc. from Smith College. She devoted her life to the chemistry of the radioactive elements.

In 1916 she was appointed docent in radiochemistry at the University of Oslo, a newly created position. During the following years she made many precise analyses of radioactive minerals and determined their age by the uranium–lead method. In 1917 she was elected to the Videnskaps–Akademi on the motion of Professor Heinrich Goldschmidt, the second woman so honored. In 1929 she succeeded Heinrich Goldschmidt as Professor of Inorganic Chemistry at the University of Oslo, the second woman professor in the university. In the 1940's she did some of the early work on the radioactivity of potassium and its significance as a heat source in the Earth. She retired in 1946 but continued her researches until her death. Throughout her life she was active in international affairs, serving on committees of the League of Nations and UNESCO. During both World Wars she worked for the relief of prisoners of war and refugees. Although V. M. Goldschmidt opposed her appointment to a professorship, he admired her greatly for her moral courage and her opposition to Naziism.

6. Hassel, Odd (1897–1981) was born in Oslo, the son of a physician. He entered the university in 1915 and took his first degree in chemistry in 1920. He then did research in Germany for some years, receiving a doctor's degree from the University of Berlin in 1924. Returning to Norway in 1925, he was appointed docent in physical chemistry and electrochemistry at the University of Oslo, and worked closely with Goldschmidt on crystal structure problems. Goldschmidt nominated him for the chemistry professorship in 1929, and when the faculty rejected this nomination Goldschmidt resigned and left Oslo for Göttingen. In 1934 Hassel was appointed to a newly established professorship in physical chemistry, which he held until his retirement in 1964. During World War II he was imprisoned for his anti–Nazi activities. He was a lifelong friend of Goldschmidt and gave his memorial address to the Videnskaps–Akademi on November 14, 1947 (see Appendix A). In 1969 he shared the Nobel Prize in Chemistry with the British chemist Derek Barton "for their contributions to the development of the concept of conformation and its application to chemistry". Hassel's activities during World War II are extensively documented in Arnold Kramish's book *The Griffin* (New York, 1986).
7. The house still stands; it has been converted into apartments.
8. *Nature*, vol. 159, p. 701, 1947.
9. Peters, Clemens was born in 1902 in Münster, Germany, and studied chemistry at the universities of Münster and Wurzburg, receiving his doctorate from Wurzburg in 1930. He then joined Goldschmidt's institute at Göttingen, where together with Mannkopff he developed the methods of quantitative determination of minor and trace elements by optical spectrography.

During the following years he was co–author with Goldschmidt on eight publications on the geochemistry of specific elements. In 1934 he joined the Badische Analin–und Sodafabrik in Ludwigshafen as a structural chemist and X–ray crystallographer, remaining there until his retirement in 1967.

10. Mannkopff, Reinhold (1894–1978) began his university studies in Freiburg, but they were interrupted by service in World War I. After the war he completed his studies at the University of Göttingen, receiving his doctor's degree in physics in 1926. In 1929 he joined Goldschmidt's institute, firstly as scientific assistant and later as docent. For geochemical problems he worked out spectral analytical methods, by developing high–dispersion prism spectrographs in the workshop of the institute. The physics of the carbon arc was the subject of his further research. He served in World War II, and after the war returned to the Göttingen institute.
11. Berman, Harry (1902–1944) was born in Boston, but his family moved to Johnstown, Pennsylvania in 1909, where he received his elementary and high school education. He then attended Carnegie Institute of Technology with interest in mathematics and engineering, but for financial reasons was forced to abandon college work at the end of one year. In 1922 he became an assistant in the mineralogy section of the Smithsonian Institution in Washington, D.C., where he acquired his intense and lasting interest in minerals. In 1924 he moved to Harvard University as assistant to Charles Palache, Professor of Mineralogy and Director of the Mineralogical Museum. Over a period of several years he completed college requirements, eventually receiving his Ph.D. degree in 1936. During the academic year 1932–33 he was awarded a scholarship for foreign travel, which he utilized in studying with J. D. Bernal in Cambridge, England, and with Goldschmidt in Göttingen. After his return to the United States he established an X–ray crystallographic laboratory in the Mineralogical Museum at Harvard. Goldschmidt admired Berman as "a very clever mineralogist," but took great umbrage when Berman published "Constitution and classification of the natural silicates" (*Am. Mineral.*, vol. 22, pp. 342–408, 1937), evidently because he failed to give what Goldschmidt considered was adequate reference to the Oslo school. During World War II Berman was active in the manufacture of quartz crystal oscillators and was killed in an airplane crash in Scotland while on a mission to Great Britain.
12. Minami, Eiiti (1899–1977) graduated in chemistry from the University of Tokyo in March, 1923. After graduation he was appointed lecturer in the Mineralogy Department, where he taught mineralogical chemistry until 1938, except for the period from

May 1933 to July 1935 which he spent as a visiting researcher with Goldschmidt in Göttingen. While there he investigated the selenium and rare earth concentrations in Japanese and European shales. His results on the abundance and distribution of the rare earth elements remained the basic data on these elements for almost 30 years; they were confirmed and extended in the 1960's by the work of Professor L. A. Haskin and his coworkers at the University of Wisconsin. Shortly after his return to Japan he was appointed Professor of Analytical Chemistry in the University of Tokyo, where he remained until retiring in March 1960, because of the age requirements of the university.

13. Strock, Lester William (1906–1982) was born of Pennsylvania Dutch parents in Chambersburg, Pennsylvania. He received his Ph.D. in chemistry from the University of Pennsylvania in 1931. He specialized in crystal chemistry, and in May 1933 arrived in Göttingen as a post-graduate guest researcher. He had been attracted to Göttingen by Goldschmidt's earlier work on crystal structures, but rapidly adapted to the new developments in geochemical research. He remained at Göttingen until November 1935, during which time he published on the geochemistry of selenium and lithium. He then spent some time in London writing a small book *Spectrum Analysis with the Carbon Arc Cathode Layer* (Adam Hilger, London, 1936) which described the technique used so successfully in Goldschmidt's institute. He rejoined Goldschmidt in Oslo from May 1937 to December 1938. From January 1939 until the early 1950's he served as research radiographer for the Saratoga Springs Commission of the N.Y. State Conservation Department. In 1951 he joined the Sylvania Electric Products Co. and worked for this firm until his retirement in 1971, investigating the crystal chemistry of compounds used in the lighting industry.

14. Witte, Helmut, born 1909, studied at the universities of Braunschweig, Munich, and Göttingen. He completed his studies in Goldschmidt's institute and received his doctor's degree with a thesis on spectral physics in 1933. After graduation he was appointed a scientific assistant in the institute, doing research on the distribution of trace elements in black shales. He worked in industry from 1939 to 1943, when he was called to the Technical University in Darmstadt, first as docent and from 1954 Professor of Physical Chemistry.

15. Noll, Walter (1907–1987) was born in Jena, and studied at the universities of Jena and Heidelberg, graduating from Jena in 1930 with a thesis on the sorption of potassium on clay minerals. In April 1930 he joined Goldschmidt's institute at Göttingen, where he investigated the synthesis of clay minerals and the geochem-

istry of strontium. From November 1932 to April 1937 he was scientific assistant in the Mineralogisch–Geologisches Institut of the Technische Hochschule in Hannover. He then joined IG–Farbenindustrie as a silicate chemist. In 1950 he was appointed docent in crystal chemistry at the University of Cologne and was promoted to professor in 1958.

16. Engelhardt, Wolf von, was born 1910 in Dorpat, Estonia, studied at the universities of Halle, Berlin, and Göttingen, and graduated from Goldschmidt's institute in 1935 with a thesis on the geochemistry of barium. He was then appointed assistant to Professor C. W. Correns at the Mineralogical Institute of the University of Rostock. After World War II he directed the geological work of the petroleum company Gewerkschaft Elwerath in Hannover. In 1957 he was appointed Professor of Mineralogy and Petrology in the University of Tübingen, a position he held until his retirement in 1978. He is noted for his researches on sedimentary petrology, on the investigation of the Ries astrobleme, and for his work on the mineralogy and petrology of the lunar rocks.
17. For a biography see: *Science and Russian Culture in an Age of Revolutions: V. I. Vernadsky and his Scientific School, 1863–1945,* by K. E. Bailes. Indiana University Press, Bloomington, Indiana, 1990.
18. The Vernadsky–Goldschmidt correspondence is in the Vernadsky papers at the Columbia University Library in New York.
19. Letter to Professor Denis Shaw, McMaster University, Hamilton, Ontario, February 6, 1973.
20. Laves, Fritz Henning (1906–1978) was born in Hanover, Germany. He studied at the universities of Innsbrück and Göttingen, and received his doctorate in crystallography from the Swiss Institute of Technology in Zürich in 1929. After graduation he joined Goldschmidt in Göttingen as assistant and docent. He was obviously a favorite of Goldschmidt, who permitted him to smoke in the laboratory, a unique privilege. During this time he studied the structure of intermetallic compounds, a major advance in the understanding of alloys. At Göttingen he was never promoted beyond docent, since the Nazi regime suspected him, a principled man, of protecting Jews (in 1933 he tried to muster the Faculty of Science at Göttingen in support of Goldschmidt). During World War II, through the influence of Paul Rosbaud, he was assigned to a group under Reichsmarschal Göring to develop an alloy "stronger than steel and lighter than air". After the war he was one of the German scientists transferred to the United States, and he joined the University of Chicago in 1948, working with Professor Julian Goldsmith on order–disorder in the feldspar minerals. In 1954 he returned to Zürich as the successor to Paul

Niggli as Professor of Mineralogy, where he remained until retirement.

21. See Appendix A.
22. Richard probably refers to Richard Wagner, composer of Parsifal and other operas.
23. *Geol. Fören. Förh.*, vol. 56, p. 385–427, 1934.
24. *Teknisk Ukeblad,* No. 11, p. 6–7, 1936.
25. Letter to Denis Shaw, August 12, 1972. Actually, Hitler came to power not as the result of a putsch in March 1933, but through the election of January 30, 1933.
26. Letter, May 6, 1933, in the archives of the University of Göttingen.
27. Letter, February 2, 1935, in the archives of the University of Göttingen.

8/Oslo: 1935–1942

1. The house is still in use (1991).
2. Saeland, Sem (1874–1940), Norwegian physicist, Rector, University of Oslo, 1927–36.
3. Schetelig, Jakob Grubbe Cock (1875–1935), the son of a ship owner, entered the University of Oslo in 1893. During his student years he was an assistant to Fridtjof Nansen, working up the oceanographic results of the "Fram" North Polar expedition (1893–96). In 1905 he was appointed a scientific assistant to Professor Brøgger, and later senior curator at the Geological Museum. In 1917 he succeeded Brøgger as Professor of Mineralogy and Petrology and Director of the Museum. His first mineralogical publication, in 1911, described the new mineral thortveitite, $Sc_2Si_2O_7$, the first mineral containing the rare element scandium as a major component. In the following years he published many papers on rare minerals in Norwegian pegmatites. After Goldschmidt's move to Göttingen in 1929 he took over direction of the Raw Materials Laboratory. He died in 1935 after a lengthy illness.
4. Archives, Geological Museum, Oslo.
5. Holtedahl, Olaf (1885–1975), Norwegian geologist, received doctorate from University of Oslo 1913, Professor of Geology and Director of the Geologisk-Paleontologisk Institut, 1920–56. Geologist on expeditions to Svalbard 1914–17, Novaya Zemlya 1921, Antarctica 1927–28; author of "Geology of Norway" (1960).
6. Bugge, Arne, brother of Carl Bugge; geologist, Geological Survey of Norway, 1921–52.
7. Bugge, Carl (1881–1968), geologist, Director of the Geological Survey of Norway, 1921–51.

8. *J. Chem. Soc (London)*, p. 655–673, 1937.
9. *Gerlands Beitr. Geophy.*, vol. 15, p. 38, 1926.
10. *Travaux du congres jubilaire Mendeleev II*. Acad. Sci. U.R.S.S., p. 387, 1937.
11. *Fra Fysikkens Verden*, vol. 3, p. 179, 1942.
12. Harwood, Henry Francis, Professor of Chemistry, Imperial College of Science & Technology, London.
13. Holmes, Arthur (1890–1965), English geologist, Professor of Geology, University of Durham, 1924–43, Edinburgh, 1943–65.
14. *Geochemische Verteilungsgesetze der Elemente*, IX, p. 34, 1938.
15. This correspondence is from the Palache papers, Mineralogical Museum, Harvard University.
16. *Am. Mineral.*, vol. 30, p. 126, 1945.
17. Kvalheim, Aslak, born 1911, was educated as a chemical engineer and joined the Raw Materials Laboratory on October 1, 1937. He was responsible for the laboratory operations and became Goldschmidt's closest collaborator. After the latter's departure in 1942, he was appointed Director of the Raw Materials Laboratory, filling that position until Goldschmidt's return in June 1946 and after his death in 1947. He was an expert in quantitative spectral analysis, and published extensively on that theme and on the utilization of Norwegian mineral resources. In 1941 he was co–author with Goldschmidt of a monograph on trace elements in meteorites, which unfortunately on account of the war was never published.
18. See Appendix A.
19. Not quite true—his mother's urn is porphyrite.
20. Archives, Geological Museum, Oslo.
21. *Norske Videnskaps–Akademi i Oslo Årbok* 1940, p. 29–41.
22. Seip, Didrik Arup (1884–1963), Professor of Linguistics, University of Oslo, 1916–1945. University Rector 1937–45, but dismissed by the German administration in September 1941, arrested by the Gestapo and imprisoned in Grini concentration camp outside Oslo; transferred to the Sachsenhausen concentration camp outside Berlin in April 1942; as a result of protests from Swedish and Finnish academic authorities he was released in December 1942, but held in civil internment in Germany until April 1945.
23. Hoel, Adolf (1879–1963), geologist and polar scientist, Professor of Polar Geography, University of Oslo, 1940. In the 1930's he became a prominent member of Quisling's Nasjonal Samling. After Seip's dismissal Hoel was appointed Prorector and in 1943 Rector of the University of Oslo. After the German defeat in 1945 he was dismissed from the university and prosecuted for his actions during the war.
24. Letter to B. Mason, September 29, 1990.

25. Suess, Hans Eduard, was born in Vienna on December 16, 1909, the son of Franz Eduard (1867–1941) and the grandson of Eduard Suess (1831–1914), both distinguished geologists who held professorships at the University of Vienna. He received his doctorate in chemistry at the University of Vienna in 1935, and was a research associate and later professor at the University of Hamburg until 1949, when he emigrated to the United States. During World War II he twice visited Goldschmidt in Oslo. He was intrigued by Goldschmidt's data on the abundances of the elements, especially the high abundances at neutron numbers 50 and 82 (later called "magic numbers"). In 1947, he published a joint paper with his colleague Hans Jensen *Zur Deutung der Goldschmidt'schen Haufigkeitsverteilung der Elemente* (The interpretation of Goldschmidt's abundance distribution of the elements) in *Naturwissenschaften,* vol. 34, p. 131. This led to the formulation of theories of nuclear structure and the origin of the elements, which resulted in the award of Nobel Prizes to Maria Goppert Meyer, Eugene Wigner, and Hans Jensen in 1963, and to William Fowler in 1983. Together with Harold Urey, Suess published a revision of Goldschmidt's abundance table in 1956. Since 1955 he has been Professor of Chemistry (now Emeritus) at the University of California, San Diego. In 1974, he received the Goldschmidt Medal, the highest award of The Geochemical Society.
26. *Applied Geochemistry,* vol. 3, p. 385, 1988.
27. See Appendix A.
28. *Applied Geochemistry,* vol. 3, p. 387, 1988.
29. Letter dated November 24, 1946; p. 63 in Adolf Hoel's book *Universitetet under okkupasjon* (Oslo, 1978).

9/Last Days—Sweden, Great Britain, Norway: 1942–1947

1. Backlund, Helge Gotrik (1878–1958), born in Estonia, educated in St. Petersburg, assistant in the Mineralogical-Geological Museum, 1902–17. Professor of Geology, Åbo Akademi, Finland, 1918–24; Uppsala University, 1924–43. He did extensive field work in Siberia, Svalbard, and Greenland.
2. Regulation 18B, effective in Britain during World War II, provided for the internment without trial of persons considered dangerous to the war effort. Sir Oswald Mosley was the leader of the British Union of Fascists.
3. *Quart. J. Geol. Soc. London,* vol. 100, p. xlv, 1945.
4. Solberg, Halvor Skappel (1895–1974), Norwegian meteorologist; Professor, University of Oslo, 1930–64.

5. Hylleraas, Egil Andersen (1898–1965), Norwegian physicist; Professor, University of Oslo, 1937–65.
6. Jensens and Lorentzens are large grocery stores in Oslo.

10/Epilogue

1. "I am aware that the new journal will, at the beginning, have to fight its way through the maze of scientific literature, but I have no doubt that the journal will soon be established and become a very valuable contribution to the development of geochemistry and cosmochemistry. I am deeply sorry that the late Professor V. M. Goldschmidt, who was an old friend of mine, is with us no longer. I have discussed the idea of such a journal with him many times after the war, and he certainly would have given every possible support to it." Letter, P. Rosbaud to F. E. Wickman, August 12, 1949.
2. The publisher was right and I was wrong. The book went through four editions and sold more than 50,000 copies in the American edition; it has been translated into German, Russian, Spanish, Portuguese, Japanese, and Malay.
3. Letter to Professor Denis Shaw, August 12, 1972. (in Russian, translated by Dr. J. Dostal, St. Mary's University, Halifax, Canada).
4. Letter to Professor Denis Shaw, June 19, 1973.
5. Letter to Professor A. A. Levinson, January 8, 1988.

Appendix A

Biographies/Memorials by P. Rosbaud, C. E. Tilley, and O. Hassel

Victor Moritz Goldschmidt (1888–1947)*

Paul Rosbaud

"The basic problem of geochemistry," said Victor Moritz Goldschmidt, "is to determine the quantitative chemical composition of the earth and to find the laws which underlie the frequency and distribution of the various elements in nature."

During the nineteenth and early twentieth centuries attempts were made, mainly by chemists, to collect the chemical and physical-chemical data relating to mineralogical and geological chemistry. In Switzerland, Schoenbein coined the name "geochemistry"; in Germany Carl Gustav Bischof published, between 1848 and 1854, his famous treatise on "Physical and Chemical Geology" which presented the chemical knowledge of his time. The great Berzelius in Sweden did much to elucidate the chemical genesis of minerals and rocks, and at the beginning of the century J. H. van't Hoff laid the foundation of the physical chemistry of salt minerals. The greatest contributions came from the work of the American petrologists F. W. Clarke and H. S. Washington. Modern geochemistry, however, had its origin in Oslo. The laws which determine the geochemical distribution of the elements were postulated by V. M. Goldschmidt from 1917 on.

Goldschmidt was born on 27 January 1888, in Zürich. His father, Heinrich Jacob Goldschmidt, was a distinguished physical chemist who came from Prague which was then, of course, part of the old Austro-Hungarian monarchy. His mother was Amelie Köhne, and Victor Moritz was their only child. In quick succession his father became professor in Amsterdam, then Heidelberg—where V.M. received his early schooling and in 1905 (*sic;* probably 1900) he was appointed professor of chemistry, succeeding Waage at the University of Oslo (then Christiania), at that time the only university in Norway. Famous for its tradition in the mathematical sciences, Oslo was also famous for its school in the earth sciences. V.M. matriculated in 1905 to study mineralogy, geology, inorganic and physical chemistry, and came under the inspiring influence of the great W. C. Brøgger, whom he soon equaled in the importance of his geological and petrological work. Except for the winter terms of 1908 and 1911 when he worked under F. Becke in Vienna and P. von Groth in Munich, he finished his studies in Oslo and obtained his doctor's degree in 1911. After two years as a lecturer he was appointed, in 1914, at the early age of 26, full professor and director of the Mineralogical Institute of the University of Oslo.

By then, shortly before the outbreak of World War I, the first scientific harvest was brought in. His earliest paper on the pyroluminescence of quartz[1] was published when he was a first-year student. The main work of this period, which has since become a classic, was his 480-page study on the contact metamorphism of the Oslo region[2] published in 1911. Throughout this monograph Goldschmidt applied physical-chemical considerations to geological problems, rightly regarding the whole planet earth as one single physicochemical system.

---

* From: *Great Chemists,* edited by Eduard Farber.

It was perhaps the first time that, by Goldschmidt's mineralogical phase rule, the applications of Nernst's heat theorem and the law of mass action on complex systems of rocks, mineralogy have come of age as an exact science.

It took Goldschmidt eight years to complete the fundamental investigations that followed his work on contact metamorphism: an intensive geological-petrographical study of regional metamorphism of the mountains of southern Norway.[3]

Long before war broke out the Goldschmidt family had become Norwegian citizens. When the general situation in the neutral countries deteriorated, and after the intensified U-boat warfare, Norway was largely cut off from overseas; sources of raw materials were badly needed. The result was that in 1917 the Norwegian government decided upon thorough research into the mineral resources of Norway, and entrusted Goldschmidt with this task. He became chairman of a Government Commission for Raw Materials and director of the Raw Materials Laboratory. This work, which was directed entirely toward practical ends during the war, was continued afterward along general lines from which developed a new branch of science, geochemistry, which from the theoretical and the practical point of view is of the greatest importance. Several schools of geochemistry existed already and a mass of material was available. Apart from the American school, pre- and postwar Russia had produced some outstanding personalities such as W. J. Vernadsky, who did pioneer work notably in biogeochemistry, i.e. the frequency and distribution of elements in plants and animals. The work of today's leading expert on the origin of primitive life, Oparin, has its roots in Vernadsky's earlier work and is mainly a geochemical and biogeochemical complex of problems. There is also the school of the other leading Russian geochemist, A. E. Fersman, who died during the last war. Unfortunately his four-volume treatise on geochemistry has never been translated. However, the fundamental laws and principles underlying frequency and distribution of elements—the basic problem of geochemistry—were developed by V. M. Goldschmidt in Oslo between 1920 and 1928 and, when he was in Göttingen, by systematic research into the geochemistry of individual elements. That Russia alone has produced some 25,000 geochemists after the war gives some idea of the importance of modern geochemistry.

With an excellent team of co-workers, and with great energy, every conceivable method of chemical and physical analysis was used for the accumulation of data for a formidable task. Analytical methods had to be checked and adopted or new ones developed; carbon arc optical spectrography was developed; chemical analysis by characteristic X-ray spectra, already used by von Hevesy and by Assar Hadding in Sweden for geochemical purposes and mineral analysis, proved very useful; plans were also made by V. M. Goldschmidt for the construction of a mass spectrometer for the determination of the mass ratio of isotopes. The outbreak of the Second World War prevented Goldschmidt from introducing this instrument to geochemical research. Today mass spectroscopy has become a widely used tool in geochemical laboratories.

One of the methods extensively used by Goldschmidt was the analysis of crystal structure by X-rays, the great discovery of Max von Laue, and of W. H. and W. L. Bragg. The use of this method broadened the original views considerably, and almost as a byproduct a new scientific discipline, crystal chemistry, was created.

One of the principal problems in the history of the earth that Goldschmidt approached is the partition of the chemical elements during the geological evolution between gas and coexisting liquid phases, the subsequent crystallization of these liquid phases, the subsequent crystallization of these liquid phases, of molten iron, iron sulfides, and fused silicates, and the distribution of the chemical constituents in these phases. According to their tendency to enter one or the other of these phases, the elements could be classified into siderophile (metal melt), chalcophile (sulfide melt) and lithophile groups (silicate melt). A fourth group, the atmophile group, includes those elements which in pregeological times differentiated into the primordial atmosphere. It is of great interest that the distribution of the chemical elements among the three phases (iron, sulfide, silicate) is closely related to their atomic volume and the structure of their electronic shells.

The main controlling factors during these stages of differentiation are the chemical affinities of the various elements toward oxygen and sulfur and their latent heat of vaporization as compared with the affinity of the most common terrestrial heavy metal, iron. Goldschmidt took as a measure of these affinities the free energy of oxidation per gram atom of oxygen of the lowest oxides of the electropositive elements and the corresponding data for their sulfur compounds. Elements which are extremely rare in the earth crust—the lithosphere—are gold, the elements of the platinum group, and also nickel, cobalt, and germanium. They can be expected in the iron phase, the siderosphere. On the other hand the alkali and alkali-earth elements, silicon, aluminum, titanium, elements with a higher free energy of oxidation than iron, will, in the primordial differentiation, have concentrated in the outer silicate crust of the earth. Incidentally, it is known that the siderophile elements—gold, platinum metals, etc.—enter fused iron and iron alloys and are thus found as by-products in iron and steel works. The other analogon in metallurgical practice is the concentration of lithophile elements in the slag.

The next step after the primordial differentiation is the cooling of our planet and the crystallization of the silicate crust, a process which is still going on in rock formation from volcanic magmas.

Goldschmidt attacked the problem of finding the general laws and principles of geochemistry "from the viewpoint of atomic physics and atomic chemistry and to find out the relationships between the geochemical distribution of the various elements and the measurable properties of their atoms and ions." He soon discovered that the principal factor regulating the entrance of atoms and ions and the distribution of the rarer elements in the crystalline phases of igneous and metamorphic rocks was the size of the atoms and ions and not their weight. During the gradual crystallization of liquid solutions, those atoms or ions of rare metals are caught in the already existing three-dimensional lattice which, because of their size, fit into this lattice. Those atoms and ions which are either too small or too big to be caught remain in the liquid. This idea led to the discovery of the fundamental relationship between crystal structure and chemical constitution:

> The structure of a crystal is conditioned by the numerical proportions, the proportion of size and the properties of polarization of its ultimate particles, and not only the structure but also other important properties of crystals, such as hardness, solubility and melting temperature, could be predicted from the properties of the atoms, and so, in the course of our geochemical investigations, the foundations of modern crystal chemistry were laid. These relationships were not limited to crystalline phases, but applied as well to the properties of solidified glasses . . .

Until then crystallography was for the mineralogist and the practical chemist: as Goldschmidt wrote, "a purely descriptive auxiliary science which made possible the recognition and distinction of crystalline materials whether they were minerals or technical products." Goldschmidt with his excellent team of co-workers—T. F. W. Barth, G. Lunde, I. Oftedal, L. Thomassen, W. H. Zachariasen and others—embarked on a hitherto unsurpassed systematic series of X-ray studies. In 1926 the atomic and ionic sizes, i.e. the radii of the greater part of the chemical elements in their different stages of electrical charge or ionization, were determined. Goldschmidt was now in a position to predict in which minerals, rocks, or ores a certain element could be found—an achievement equally important scientifically and practically. To give only one example from innumerable cases which has been proved: nickel and magnesium have as divalent positive ions the same radius of 0.78 Å; this is the reason why nickel is found in magnesium silicates of igneous rocks, and the percentages of both follow each other closely. Goldschmidt could soon predict the sequence of fractionated crystallization, taking into account the electrostatic forces between the positive metal ions and the negative silicate ions.

However, there are additional forces which support the electrostatical attraction of the ions in the crystal frame work, i.e. van der Waals forces, polarization forces, and the electron pair forces. They result in a measurable contraction of the lattice which is highest in the case of nickel orthosilicate, followed by magnesium and then by iron (ionic radius = 0.83 Å). Nickel, in the fractionated crystallization from magmatic or other solutions, is the first, and iron the last to enter the mixed crystal. The concentration of nickel in magnesium-orthosilicate is sometimes of economic importance.

Goldschmidt could further predict the sequence of crystallization in cases of elements of the same size and practically the same ionic radius, but different electrical charge. If a divalent element, such as magnesium, is the main constituent in a crystal in contact with a saturated solution containing, as minor constituents, a trivalent element, such as scandium, and a monovalent, such as lithium, then, owing to the excess of electrostatic attraction, the trivalent scandium is preferentially built into the crystal—"captured"—whereas the univalent element lithium, owing to its smaller electrostatic charge, is only "admitted" to the lattice toward the end of the crystallization period. We understand now that scandium is not nearly as rare as it was formerly considered, but is a frequent constituent of many igneous rocks, especially magnesium silicates.

The problem of ionic substitution in minerals, which has occupied the mind of every mineralogist, was simply and elegantly solved by Goldschmidt's fundamental laws of crystal chemistry.

The chemical and geochemical behavior of elements with the same valency and almost the same ionic radii is of especial interest. The classical examples are the pairs aluminum/gallium and zirconium/hafnium. The trivalent gallium (radius = 0.57 Å) is "hidden" in aluminum (radius = 0.62 Å) and practically every aluminum mineral, every bauxite contains up to 100 g of gallium per ton of aluminum. It accompanies aluminum through all the stages of industrial manufacture, and can be detected in practically every aluminum alloy. The tetravalents zirconium and hafnium have the ionic radii Zr = 0.87 Å; Hf = 0.86 Å. It will be remembered that in zirconium minerals von Hevesy discovered hafnium, the existence of which was predicted by Bohr's atomic theory. Goldschmidt called this hiding of a rare element in another "camouflage." The most famous example of camouflage is found in the rare earth elements, and the celebrated elucidation of the geochemical and chemical behavior of this fascinating group of elements, closely related to their ionic radii and valency, is one of the most magnificent achievements of Goldschmidt's work.

In his monumental work *Geochemische Verteilungsgesetze der Elemente,* I–IX,[4] Goldschmidt laid down the fundamental laws of geochemistry and crystal chemistry. These 600 pages contain the life work of one who, more than anyone else, has laid in two branches of science the foundations on which later generations will build. Nos I–VIII contain the result of his work in Oslo, IX (the "ninth symphony" as he used to call it) originated from one of his happiest periods of life in Göttingen. No. VII contains the famous laws of crystal chemistry and No. VIII the relations between the structure and properties of crystals. One of the problems Goldschmidt approached in the latter communication are the factors influencing the hardness of crystals. Based on earlier work by Reis and Zimmermann, and by Friedrich, Goldschmidt systematically—supported by hundreds of investigated substances—extended our knowledge by showing how the hardness of a crystal was determined by the interatomic distance, the valency, and the structure type. Goldschmidt, however, did not stop here and went on to the question of whether it is possible to "construct" crystals of certain well-defined and desired properties, such as hardness, melting point, solubility, optical refraction, chemical reactivity, etc. Unfortunately this work on "model structures" of crystallized silicates and silicate glasses has remained comparatively unknown in the English-speaking world, as it was published in a journal not easily accessible (*Zeitschr. für technische Physik,* 1927). In this paper, readers will find the scientific basis of a problem of quite fundamental importance—how to construct solid materials of certain physical, chemical, and mechanical properties.

The year 1929 was a decisive year for V.M. (as he was affectionately called). In that year he was invited to join the faculty of natural sciences of Göttingen as full professor and head of the Mineralogical Institute. Other invitations from foreign universities had come to him, and more than one university was anxious to obtain the collaboration of the recognized leader in his field. Stockholm had tried previously; V.M. decided in favor of Oslo. A few years before he went to Göttingen the University of Munich had planned to appoint V.M. to the vacant chair of Paul von Groth. The failure of this proposal is a typical and ominous sign of the political influence of a small group of professors in the town which later became the hotbed of Nazism. Unfortunately at the decisive meeting of the faculty a number of politically moderate and reliable people were absent, believing that Goldschmidt's appointment was already decided. As a result the invitation was turned down because "We already have one Jew at the Faculty." In protest against this racial discrimination, the "Jew at the Faculty" resigned. He was Richard Willstätter, Nobel Laureate, and one of the greatest organic chemists of his time.

In view of this experience, V.M. may have been reluctant to accept an invitation from another German university. However, Göttingen was not Munich and, in the tradition of its great permanent Under-Secretary Althoff, the Prussian Minister of Education, treated V.M. with the greatest courtesy and, fully aware of the great importance of this appointment, agreed to all V.M.'s wishes. It is said that after a meeting for Goldschmidt's official acceptance and final decisions the very dignified Under-Secretary and his advisers joined in a dance of joy.

Goldschmidt, with his father and their Norwegian housekeeper, Frökken Marie Brendigen (his mother had died shortly before), moved to Göttingen where V.M. bought a house in Wagnerstrasse. On arriving in Göttingen, father and son found the town "dressed over-all"—not to greet the distinguished scientist, but to honor Colonel Düsterberg, the leader of a somewhat nationalistic organization of soldiers called "Stahlhelm" (the Steel Helmet). The Stahlhelm was harmless compared with those organizations that replaced it after 1933; Colonel Düsterberg, being only "partly" Aryan, was forced by the Nazis and S.S. to resign. Nevertheless, it gave the Goldschmidts an ominous foretaste of things to come.

There was a galaxy of eminent men of science in Göttingen at that time. The mathematical tradition—uninterrupted from the days of Gauss—was carried on by David Hilbert, Courant, Landau, and others; the organic chemist and Nobel Laureate Wallach was living in retirement and his successor (also a Nobel Laureate), A. Windaus, of cholesterol fame, was at the height of his career. Another Nobel Prize winner was Zsigmondy; the physics department was headed by the triumvirate James Franck, Max Born, and W. Pohl—after 1933 the first two, both Nobel Laureates, became "unbearable." Prandtl, one of the founders of modern aerodynamics, was one of Göttingen's great men; so was G. Tammann, the physical chemist who was still very active after his retirement, and who was succeeded by A. Eucken. The zoologist Kühn and the astronomer Kienle became great personal friends of the Goldschmidts. There was also a young generation, many of whom have since become world-famous.

The few years before 1933 were perhaps the happiest in Goldschmidt's life. In Oslo he was something of a lone wolf; in Göttingen he not only inspired his own students and colleagues, but drew inspiration from them all—the physical chemists, the inorganic chemists, the physicists, astronomers, and biologists. It was an almost ideal cooperation among scientists of every branch, and nobody sat in an ivory tower, anxiously guarding the results of his own work from intruders. In addition there was a constant exchange of ideas with foreign visitors and guests.

The Mineralogical Institute was housed in a former school which was adapted according to V.M.'s requirements. One wing contained the lecture theater, students' working rooms and laboratories, and the teaching collections of crystals, minerals, and rocks. The teaching department was connected to the third-floor research wing by a wide corridor in which magnificent specimens of minerals were displayed. There, a group of young and very able scientists set out under the direction and guidance of V.M. to study the geochemistry of individual elements. The laws which govern the distribution of elements were postulated in Oslo. They were now applied to the systematic survey of a large number of elements.

This period between 1929 and 1935 or, to be more precise, 1933, was the happiest epoch in the lives of V.M. and his aging father H.J. There the lonely man found not only devotion of his collaborators, which he had enjoyed in Oslo, but also the warm friendship and admiration of his colleagues.

In those pre-war days Goldschmidt was a heavily built, stout man, shy and often difficult to approach. He had an almost old-fashioned courtesy which could change to an outspoken but equally courteous frankness in scientific and other arguments. Shortly before he died he told me a story which is a clue to his strange personality. When his father moved to Heidelberg young V.M., who was about six years old, attended a private school. His mother had impressed upon him that he must always be polite and also tell the truth, however disagreeable. After a few days at his new school, the headmaster, a big man with a black beard and of alarming appearance, asked young V.M. how he liked the school, and ended by saying: "I hope we will become good friends. Do you already like me?" Aware of his mother's admonition, after a moment's pause V.M. replied: "Nicht so besonders, Herr Göckel" (not particularly, Sir).

As well as this "truthful politeness" which is found in some of his polemics and in many of the discussions after scientific meetings (not always to the delight of those concerned), he had a grim and sarcastic sense of humor, and was very quick-witted. In all scientific and personal affairs he was completely sincere, honest, and incorruptible. There was no room in his department for anyone who did not measure up to his own scientific and moral standards, nor for sloppy and untidy workers. He was enthusiastic about new, exciting, scientific results wherever they came from; but his criticism of unsound or too highly publicized work could be devastating. He had a certain naive streak of vanity, and he was easily hurt if insufficient credit was given to his work. I well remember his reaction to a long review article on some fundamental problems of geochemistry by a well-known chemist, in which only one reference was given to a paper by V.M., and that, on a minor subject. However, this was after 1933 and the author had undergone a color metamorphosis to brown—the Nazi color. It is only fair to emphasize that in the case of V.M.G. this was one of the rare exceptions. Almost all of his collaborators and colleagues stood by him in the dark days after 1933. Only one case is reported in which a younger and presumably ardent Nazi scientist dared to refer to V.M.'s name in the Nazi prescribed form—V.M. (Israel) Goldschmidt—in the Nazi-sponsored *Zeitschrift für die gesamte Naturforschung* (in the Germany of those days nicknamed "Zeitschrift *gegen* die gesamte Naturforschung").

He was by nature a pessimist and, partly perhaps because of his Jewish origin and of the anti-semitism which he had experienced from his childhood on, he occasionally suffered from a feeling of persecution. This was why he did not readily trust people until he knew them well and also why he was so responsive to genuine overtures of friendship. His kindness and generosity were unfailing, even to people unknown to him personally. He was a real animal lover and his garden was almost a bird sanctuary. There is the story about his learning the cat language, an attempt which came to an abrupt stop when he was attacked by a large tomcat to whom he had, apparently, unwittingly been extremely rude. The household, in addition to its human inhabitants, consisted of the famous dachshund "Bazi"—which tyrannized father and son; three squirrels demanded and received a home in the bathroom, a toad hibernated in the cellar, and a family of bats resided in the loft. Each had a name, and the toad shared his with a well-known and not too endearing character in Göttingen. One of the squirrels was named Parsifal ("because he was indescribably stupid") another was Richard ("Weil er so stark riecht"), but the most notorious was Magdalena, who came to a sad and untimely end. V.M. gave a birthday party for Bazi and several dog guests were invited to share the birthday cake. In the excitement, Bazi concluded that on this special day everything was permitted and ended the feast by devouring Magdalena.

Only a few of his devoted assistants and collaborators in Göttingen can be mentioned—most of them are today heads of university departments: F. Laves, Zurich; Th. Ernst, Erlangen; H. Hauptmann, São Paolo; W. v. Engelhardt, Tübingen; H. Witte, Darmstadt; R. Mannkopff, Göttingen; C. Peters, Vienna.

The scientific harvest of this short Göttingen period was particularly rich. From 1930 to 1935 Goldschmidt published an inspiring series of papers on the geochemistry of rarer elements. In each case terrestrial material from all parts of the world was analyzed; meteoritic matter was also included. New analytical methods were developed, such as the cathode layer technique which proved to be one of the most valuable methods for the detection of elements dozen to and below 0.01%. It is impossible to condense into a few paragraphs the results of well over two dozen papers, dealing with the geochemistry of the alkali metals, nickel, the noble metals, germanium, gallium, scandium, boron, arsenic, and selenium. In almost every case the abundance of the elements in the earth's crust, in meteorites, and in stellar atmospheres were investigated. One typical example is the investigation by V.M. of the geochemistry of germanium and its occurrence—with other rare elements—in coal and coal ashes.

This element is distributed in silicates of the upper lithosphere, owing to the similarity of the ionic radii of Ge and Si. It is also present in iron meteorites, and Goldschmidt found striking concentrations of it in the ashes of certain coals. A preliminary spectrographic analysis of the ashes of the boiler in V.M.'s institute and the soot in the chimney led to a systematic investigation of the geochemistry of coal ash. The highest amount—1.6% GeO_2—was found by V.M. in ash from samples of Northumberland coal. Still higher amounts (up to 9%) were found later in lignites from the District of Columbia, and in certain Russian coals.

The biochemical, or rather biophysical, mechanism of the enrichment of this and other rare elements—boron, arsenic, beryllium, cobalt, nickel, platinum metals—is still far from clear and needs further elucidation. Among those elements most highly concentrated in coal ash is arsenic. V.M. thought that one of the causes of the professional disease of chimney sweeps, skin cancer, may be the high content of arsenic in soot and flue dusts. It should be noted that petroleum shows a still greater affinity for trace elements than coal, and certain ashes of petroleum are used as a vanadium ore containing up to 70% V_2O_5.

Some thirty years ago the metallurgy and the industrial uses of minor elements were practically unknown. It did not "pay" to recover rare elements or to develop industrial uses for these. Today, thirteen years after Goldschmidt's death, some of the minor elements have become of the utmost industrial importance. Goldschmidt showed where to find them and laid the foundation of their geochemical and crystal-chemical behavior.

The peaceful lives of Goldschmidt, father and son, were fundamentally changed and their work interrupted in 1933. The "thousand years" of the Third Reich had begun. Monsieur François Poncet, French Ambassador in Berlin, summed it up "C'est la victoire des Boches sur les Allemands." Göttingen and the other German centers of learning were systematically destroyed by the Nazis. Many eminent scientists deeply regretted this—but they did not protest, or make a firm stand against the measures taken by the Nazis against science and the arts. Some of V.M.'s students went to Berlin in the hope of convincing the Ministry of Education that V.M. must remain in Göttingen. Goldschmidt and his father, who had not previously belonged to any faith, ostentatiously joined the small Jewish community in Göttingen. V.M. was determined, in the interests of his students, to continue his work at the University as long as he could, knowing that soon he too would be dismissed and forced to leave the country. He became more lonely and withdrawn than before, constantly careful not to "compromise" or endanger his non-Jewish friends and colleagues. His sense of humor became more grim than ever. His father was once perturbed at being greeted by someone with "Heil Hitler." Goldschmidt retaliated by remarking that surely no atheist could be offended if greeted by "Grüss Gott".

He continued to work on his extensive research program; early each morning and again until late each night he sat at his ancient double-keyboard typewriter, writing scores of letters to colleagues and friends abroad, recommending young scientists who had lost their positions for racial reasons. Many young scientists owe to V.M. not only their careers in countries outside Germany, but even their lives.

In 1935, after a new wave of anti-Semitism, Goldschmidt resigned his chair and returned to Norway. The Norwegian citizenship which he had to resign when he went to Göttingen was immediately granted and a chair at Oslo University offered to him. The Goldschmidts arrived back in Norway almost penniless; although emigrants and refugees from Germany were allowed to take their personal belongings with them, the money allowed was only ten marks per person. Much later part of Goldschmidt's money was transferred to Oslo as the result of a very courageous and outspoken letter written by F. K. Drescher-Kaden, V.M.'s successor in Göttingen, to the president of the German Reichsbank, Hjalmar Schacht. The house in Wagnerstrasse and all his other possessions were lost without compensation from Germany.

He rented the top floor of a small house in Holmenkollen, and divided all his energies between his scientific work and helping refugees from Germany and the countries invaded by the Germans before the war. He wrote the ninth and final publication of the "Verteilungsgesetze," a book of 150 pages covering the abundance and distribution of the individual elements and atomic species, and geochemical and cosmochemical data; in this attempt to estimate the mineral mass ratios of the chemical elements he also included isotopes, thus laying the foundation stone of isotope geology. He resumed his work for the Norwegian Raw Materials Laboratory, assisted by his devoted collaborator Aslak Kvalheim (who later became his successor in this governmental laboratory), and he continued his geochemical research. As a side line, and a money-making hobby, he resumed the industrial work he had done after World War I, when he did so much to develop the utilization of Norwegian raw materials. On the west coast of Norway there are large deposits of olivine; Goldschmidt discovered that it has refractory properties, and he himself developed the methods successfully used in industrial furnaces and ceramic kilns of transforming olivine rock into industrial refractories. In addition he made use of olivine sand and crushed dunite rocks instead of quartz sand for making molds for foundries. The olivine molds not only produce excellent castings of steel and high-melting ferro-alloys, but also reduce, and may even eliminate, the danger of silicosis in foundries.

In the tranquility of his new home Goldschmidt was at peace again, although he was distressed by tension and oppression in Germany and anxious about many of his friends still living there. His correspondence was greater than ever, and the file marked "N.A." (non-Aryan) grew daily. A personal tragedy for him was the sudden death of his father, and he became more lonely and withdrawn than before. Future and worse developments in the political theatre, however, soon made him realize that death had spared his father from participation in the coming cataclysm.

I went to stay with him in Oslo shortly after his father's death. He had aged, his health had deteriorated, and he was embittered and tired. But he went on with his scientific and his relief work with almost inexhaustible energy. He spent more than he could afford on helping other people and managed to bring an aged aunt and a niece—still a child—from Germany to Oslo, and the niece from there to the U.S.A. He still enjoyed his food and an occasional drink and he maintained his old-fashioned courtesy. His humor had become more whimsical, more grim, and sometimes even caustic. I remember one day when I asked him to come with me to take some flowers to the crematorium as a tribute to his parents. We stood reverently in front of the urns, all made of beautiful green Norwegian olivine, two with the ashes of his parents and a third, empty and destined to hold his own, and finally V.M. remarked dryly: "Ja, ja, the whole family in magnesium orthosilicate."

His pessimism in those pre-war days was sometimes almost unbearable. He knew the omens and foresaw the coming pandemonium from which there was no escape, and yet he worked and made plans for further research. The spectrographic unit was in full swing, and plans were ready for a new mass spectrometer with which to study the abundance and distribution of isotopes. And then war broke out, bringing the darkest days of Goldschmidt's life. He had always been a man of strong likes and dislikes and could not always see eye-to-eye with his colleagues. During the war, and especially when Norway was overrun by the Germans, he forgot all his animosities in a personal attempt to resist the waves of evil which swept over Europe. He kept in touch with those friends in Germany who still dared to write to him and, in order not to compromise them during censorship, signed his letters illegibly. His main correspondence was with English friends; even during the Nazi occupation many messages reached the Allies through Goldschmidt's help.

By nature timid and anxious, he grew in stature during the war years, but his health suffered. Finally he was arrested by the S.S., his property was confiscated and he was sent to one of the concentration camps which the new masters had installed. These camps were full of the Norwegian intelligentsia—university teachers, lawyers, writers, etc.—regarded as dangerous by the regime. It is true, and it should never be forgotten, that nearly everyone in Norway was engaged in the resistance. Goldschmidt's case, however, was a special one: in addition to being one of the most prominent Norwegians, he was also a Jew, and therefore earmarked for extermination in the gas chamber. In November, 1942, he was listed for deportation to Poland, but at the last moment, probably through the clever intervention of the Norwegian police, and owing to his poor health, he was temporarily released. Then, with the help of the Norwegian resistance he was smuggled across the Swedish frontier in a load of hay which some German soldiers prodded with a fork.

Goldschmidt did not stay long in Sweden. He was eager to offer his services to the Allies, and in the spring of 1943 he was brought to England. It was in this country, and especially in Scotland, that, although seriously ill, he felt at peace again. Under the auspices of the Agricultural Research Council he first went to Craigiebuckler, near Aberdeen, to work at the Macaulay Institute for Soil Research. He was warmly received by its director, Dr (now Sir) William Ogg and his wife, and for the first time in many years he knew what it meant to live in a free country. He took a great interest in soil research and, when Ogg became director of Rothamsted Experimental Station, he moved with him to Harpenden. He was liked by everybody and soon affectionately called "Goldie"; he took the greatest interest in the Institute's work on trace element distribution, and used to remark that it was in England, and through his contacts with Dr Ogg, Dr Alex Muir, and others that he began "to understand what geochemistry really is"—put with great modesty, this was certainly a great appreciation of the excellent work of British soil chemists.

Goldschmidt always hated the writing of papers

and books, and he used to tell his pupils in Göttingen to write their papers as concisely as a Scotsman would if they were to be sent by telegram. It was with difficulty that Dr Ogg and his colleagues eventually persuaded V.M. to write his last *magnum opus,* and gave him every facility for his work. But the task was too great and his health already too seriously undermined. Seven hundred folios were written, but he could not finish the rest. The ultimate book *Geochemistry* was completed after years of hard work by Alex Muir with the collaboration of A. Kvalheim and published posthumously by O.U.P. in 1954.

Many honors came to Goldschmidt—he received an honorary degree at Aberdeen University, and also the highest award of the Geological Society, the Wollaston Medal; but nothing pleased him more than his election, as one of its fifty foreign members, to the Royal Society, London. Perhaps, had he lived longer, he would have received a Nobel Prize, which he so richly deserved. His deepest satisfaction, however, was the kindness he received, and the fact of his freedom—in Göttingen, when one Nazi riot followed another, he and his father carried hydrocyanic acid in a capsule for emergency use as a final evasion of Nazi cruelty. In Norway, too, during the occupation, he carried poison in his pocket. On one occasion a university colleague of his in Oslo asked Goldschmidt for a capsule. "This poison," answered V.M., "is for professors of chemistry only. You, as professor of mechanics will have to use the rope." This was the typical grim humor of Goldschmidt. In England he did not need the poison any longer, and in a gay ceremony at Rothamsted it was solemnly buried deep in the garden.

I saw Goldschmidt again in January, 1946, after a long interval, and found him greatly changed in appearance. From then on I went to Harpenden once or twice every week to spend the day with him. His agonizing experiences in Norway and his recent illnesses had made him lose weight and he looked more like a pathetic survivor of a Nazi concentration camp. In spirit, however, he was indestructible; he was still easily hurt if somebody failed to give credit to his work and he suffered, often with justification, from a sense of persecution. His English, although grammatically correct and even fluent, was known as "Goldschmidt English" because of his quaint pronunciation. As usual he was more generous than his means really allowed and the pockets of his suits, which always became baggy, bulged with dozens of small pieces of paper: notes of atomic constants, old bills, bus tickets, and the addresses of people to whom he wished to write, or whom he wished to help in their academic career.

In one fundamental way he had changed: the bitterness had gone—he no longer hated those who, in so grim a way, had twice interrupted his work, undermined his health, and shortened his life. One episode influenced V.M. deeply and occupied his mind until the end of his life: it happened early in November, 1942, in the County Hospital in Tönsberg, Norway, which was then the internment camp for Norwegian Jews. After a day of humiliation and torment by his Nazi jailers, V.M. talked to two other prisoners, whose names deserve to be recorded: Moses Katz, an orthodox Jew and a hosiery peddler, and Lesser Rosenblum, socialist, atheist, and manufacturer of umbrella handles. V.M. suggested that they should remember the names of their tormentors, so that any survivors might exact retribution. The reply of the pious Moses Katz was a surprise to V.M.: "Revenge is not for us; that must be left to the Almighty." With the arrogance of a scientist confident of his superior knowledge, V.M. asked what prayers would be permissible to God from men in their position. Katz replied without hesitating a moment: "You may pray that the hearts of your enemies may be enlightened." Goldschmidt, still not admitting defeat, turned to the atheist Rosenblum and asked for his view. His reproof was equally unexpected: "We must break the evil circle of retribution, or there can never be an end to evil."

Goldschmidt became very humble after this experience. He had escaped, but his two friends from Tönsberg were facing death in Poland's gas chambers. He regarded their sayings as lucid and practical improvements on the Old Testament. Through them he learned not to forget, but to forgive.

The prolonged strain he had undergone, his escape, and the frequent adjustments to new countries and languages and people had completely undermined V.M.'s health, and for weeks he had to stay in bed. On one occasion he was staying with the Oggs. Dr Ogg, on his way to bed very late, heard groans coming from V.M.'s bedroom. He had had a heart attack and was convinced that he was dying. Between groans he kept saying "This is the end. I have done my best for Norway and Britain." A doctor was called at once and Goldschmidt was given oxygen; but his condition was critical and the doctor, scarcely expecting him to live until the morning, gave Dr Ogg the name and telephone number of an undertaker. V.M. appeared to be unconscious but evidently heard these remarks. He had enormous will power and was not yet prepared to be defeated by illness. When he was being carried out to the ambulance in the morning he was joking about the undertaker.

His mind worked constantly; many fascinating theories occupied him and were developed in talks with visiting friends. No record of these is available as Goldschmidt never committed to paper, or published, ideas which he could not yet prove. He strongly disliked the premature publication of theories: "Make two thousand experiments, and then you have your theory and can publish both," he used to say.

However, he did write two papers which, for different reasons, he was unwilling to publish. There exists a handwritten manuscript, "Geochemical Aspects of the Origin of Complex Organic Molecules on the Earth as Precursors to Organic Life." It was published after his death in a series of popular scientific books but has remained practically unknown to scientists, and Goldschmidt himself would never have agreed to its publication in its unfinished form. Some of the ideas expressed in this draft may one day reappear and their significance be recognized.

The second paper—actually more a note—was written in cooperation with G. Nagelschmidt and was never published because Goldschmidt felt it was, perhaps, undignified to end his fundamental scientific work with a note on *album Graecum* (dog feces), a prosaic and unattractive product, however interesting in its chemistry, in which insoluble phosphates are transformed into white and soluble ones. It may be recalled that before World War I it was used in the manufacture of fine, high-grade leather. The idea of this note must have come to V.M. during the war, when he grew vegetables in a small suburban garden outside Oslo where many people kept dogs. *Album Graecum* is an ideal phosphate fertilizer for a poor soil; but if it would be undignified as a scientific paper V.M. thought it would be even more undignified for a middle-aged professor to collect *album Graecum* from suburban pavements for his garden. He was in any case a very reluctant gardener, and had a widely representative collection of weeds in his allotment.

For several months in Harpenden, when he was too weak to leave his bed, he must have felt that his life was slowly drawing toward its close. He did not speak of the disappointments, the sorrows, the pains, and the dangers he had experienced, but of bygone and happy days; of Göttingen and his friends and pupils who resumed contact with him after the war. He was happy that his old Institute had two of his friends as his successors, first F. K. Drescher-Kaden and after him C. W. Correns; and he was happy that his work was carried on by the younger generation which had been so strongly influenced by him. In his weakness V.M., who had never spoken German since 1939, sometimes lapsed unconsciously into his mother tongue. It was a moving experience to visit Goldschmidt in those days—the lonely man with some enemies, but with many devoted friends. These he remembered in gratitude, the others he forgot. "Should you see Dr Lonsdale or Professor Tilley, and if you write to my old friend Eskola, please remember me to them," he would say, wanting them all to know that he was still alive. He could no longer go to see the Oggs, but they came and looked after him and he kept his friendship and gratitude for them until he died.

However, his time had not yet come. He recovered sufficiently to prepare for his return to Oslo. Once he felt better, his unfailing will power took possession of him, and he made plans to resume his scientific work on a large scale. At the end of June, 1946, he left England and arrived safely in Oslo. "Father's Return" was celebrated enthusiastically at the Raw Material Research Laboratory, and he was delighted to see among other friends W. Zachariasen of Chicago University, one of the pupils of whom he was most proud. At his flat much of his furniture had been recovered with the help of the government, and there were letters from Göttingen, offering him his old chair and from China, inviting him to organize raw materials research in that country. Soon, however, he had to undergo examination and treatment in several hospitals. His reports were optimistic, heart, kidneys, and lungs seemed to be satisfactory, and he hoped to have several years ahead in which he could work, provided he "followed the recognized principles of Moses Katz." Miss Brendigen had again joined his household, and had provided him with what he regarded as a delicious specialty, fresh whale meat! In October for the first time he mentioned a black spot on his leg; he received X-ray treatment, but the growth proved to be malignant and required surgery. The carcinoma could not be removed completely, and again Goldschmidt had to go into the hospital. Still he continued working on his book, directing scientific and industrial research and writing letters to his friends. In March, 1947, he entered the surgical ward for the sixth operation. On March 19 I had a few scribbled lines that the operation had been successful and that he would be allowed to go home on the following morning. He arrived home on the twentieth, complained of a sudden intense pain in his head, and died almost immediately.

In one of his last letters to me he wrote:

> The wisdom of the Moses Katz principles is undeniable. . . . And I am fully convinced that it is my duty towards science and decency to stand firm in continuing my work as long as health permits, thus giving an example to at least some of my junior colleagues. Often I think that (to maintain these principles) to be even more important than my contributions to scientific and industrial research and my scientific teaching. To set a new standard of morality is a matter of great urgency in these times . . .

NOTES AND REFERENCES

1. Die Pyroluminescenz des Quarzes. *Kgl: Norske videns.-skab. Selskabs. Skrifter,* No. 5, 1–15 (1906).
2. Die Kontaktmetamorphose im Kristiania-Gebiete, *ibid.,* No. 1, 1–483 (1911).
3. Geologisch-petrographische Studien im Hochgebirge des suedlichen Norwegens. I–V, *ibid.* (1912, 1915, 1916, 1920).
4. Geochemische Verteilungsgesetze der Elemente. I–IX, *ibid.* (1923–1937).

Acknowledgements—I gratefully acknowledge the help and advice of friends of V. M. Goldschmidt: Professor C. W. Correns, Göttingen; Dr Alex Muir, Harpenden; Sir William Ogg, Edzell, Angus; Professor C. E. Tilley, F.R.S., Cambridge.

Victor Moritz Goldschmidt (1888–1947)*

C. E. TILLEY

VICTOR MORITZ GOLDSCHMIDT was born in Zürich on 27 January 1888, the only child of Heinrich Jacob Goldschmidt and Amelie Goldschmidt (*née* Koehne). His earliest schooling he received at Amsterdam and Heidelberg, and later at Oslo, where his father became professor of chemistry in the University in 1900.

In Oslo he attended the Vestheim school, matriculating in 1905, and in the same year he and his parents became Norwegian citizens. At the University which he entered the following session he studied mineralogy and geology under W. C. Brøgger, and inorganic and physical chemistry under H. Goldschmidt and Th. Hiortdahl. There, through the influence and inspiring personality of Brøgger, he was soon to be drawn into research in the fields of mineralogy and geology, and to embark upon the first of the remarkable series of researches which have illuminated the study of rock metamorphism.

In the course of these formative years he studied for a short period during the winter of 1908–1909 under F. Becke in Vienna, and he spent the winter session of 1911–1912 with von Groth at Münich. He gained his doctorate at Oslo in 1911, was made docent in 1912 and appointed in 1914, at the early age of 26, Professor and Director of the Mineralogical Institute in the University. From this time till 1942, apart from a period of six years at Göttingen (1929–1935), his interests and activities were to be associated with his chair at Oslo.

Goldschmidt's achievements in mineralogy and petrology, and in the study of the regional geology of Norway, rank in importance with those of his illustrious teacher and predecessor. They were sustained and succeeded by researches of brilliant distinction which not only laid the foundation of a new science of inorganic crystal chemistry but gave to geochemistry its modern orientation and much of the impulse to its present phenomenal development. Pre-eminent in achievement, of great range and depth, these collected investigations remain to give him rank as one of the most outstanding men of science of his day and generation.

At Oslo and Göttingen he gathered around him an able and enthusiastic group of pupils drawn from many countries, attracted and inspired by his genius and example. Many of them, now come to advancement in professorships or high posts in industry in Europe and America, are building on the foundations he laid and have been responsible for some of the most recent developments in the wide fields he made his own.

Goldschmidt's first published paper appeared in his first year at the University. It dealt with the pyroluminescence of quartz, a subject which had taken his youthful interest as early as 1904. This earliest effort in research gave clear indication of the resourcefulness and thoroughness with which later investigations were to be conducted.

He contributed a paper on the germanium mineral argyrodite, and indicated how the measurement of radioactivity of minerals could be used as an aid in the identification of some of the rarer species, such as orangite, samarskite, euxenite and polycrase. But the main work of these early years was to lead to his doctorate thesis on contact metamorphism in the Oslo area.

In the spring of 1907 he had begun a study of the mineral deposit of Årvollsdalen near Oslo with its rich assemblage developed in the calc silicate xenoliths in the surrounding nordmarkite. Eventually at the suggestion of Brøgger these studies were enlarged to embrace the aureoles of metamorphism around the major igneous intrusions of the Oslo area. The Permian plutonic igneous masses of the region had long been celebrated by the work of Brøgger, and Goldschmidt was now to engage on a close investigation of the thermal metamorphism induced at their contacts with the varied group of Palaeozoic sediments. As a result, there appeared in 1911 in a memoir of more than 400 pages a masterly contribution to rock metamorphism which has since become a classic.

* From: *Obituary Notices of Fellows of The Royal Society*, Vol. VI, 1948–1949, pp. 51–66. Reproduced by kind permission of The Royal Society.

A study marking a major advance in the correlation of the chemical and mineralogical composition of thermally reconstituted rocks, it contained the first successful essay towards a systematic classification of rock mineral assemblages in the light of phase rule theory.

Goldschmidt's mineralogical phase rule was developed by the consideration of solid phases in contact with their saturated solutions, a method having the advantage that it probably provides a real picture of the processes in operation in contact metamorphism. In this way he set up for the shale-limestone group of sediments an orderly group of ten equilibrium assemblages (classes) the validity of which has been abundantly demonstrated in later work.

Goldschmidt was among the first in Europe to recognize the importance of metamorphism in connexion with skarn development associated with certain classes of ore deposits. Thus, in this same study he recognized in the Oslo region a pneumatolytic metamorphism associated with the ores of the contact zone, responsible for the production of the associated skarns rich in hedenbergite and andradite. He thus joined forces with Kemp and Lindgren in America in recognizing the importance of magmatic transfer of common elements of ore deposits, iron, silicon, sulphur and copper, especially in limestone environments. Later he was to provide in the Langesundsfjord area of the province, clear evidence of sodium metasomatism associated with nepheline syenite pegmatites at their contacts with Downtonian sandstone.

His physico-chemical approach to problems of rock metamorphism led further to an interesting attack on the important problem of geologic thermometry. Assuming the validity of the Nernst heat theorem he endeavoured to calculate the *pt* curve of the univariant equilibrium $CaCO_3 + SiO_2 \rightleftarrows CaSiO_3 + CO_2$. The data for this evaluation were uncertain, as he recognized, and he drew attention to the value of a precise experimental investigation of the system in providing a reliable temperature recorder in metamorphism. The system still awaits experimental study.

The winter of 1911–12 spent at Münich with Groth provided at this time two papers, one an investigation on the change of interfacial angles of non-cubic crystals at low temperatures, research which in conjunction with the work of F. E. Wright on the same materials at high temperatures, has provided a remarkably complete temperature history of two such important substances as quartz and calcite; the other an account of his ingenious conversion of a single circle goniometer to a two circle type which has all the advantages of the more costly three circle instrument in permitting, in a single setting, ready measurement and comparison of interfacial and interzonal angles of isomorphous series of substances.

With the completion of his studies on the contact metamorphism of the central part of the Oslo region Goldschmidt turned his attention to problems of regional metamorphism. As a first stage in this approach he had intended a comprehensive investigation of the metamorphism of the Langesund region where Brøgger's studies had demonstrated that through combined action of heat and the local impress of stress, rocks had developed with the habit of crystalline schists. This part of his programme though apparently brought near to completion was never published.

For the main field of his labours in regional metamorphism he chose the great Caledonian geosyncline of southern Norway where a sedimentary succession can be traced through all stages of alteration into crystalline schists, their metamorphism being in part associated with a contemporaneous suite of eruptives. The geological work of the pioneers, Brøgger, Bjørlykke, Reusch and Rekstad had provided the basis from which an intensive geological-petrographical study might now be elaborated. The region over which his projected studies were to range comprised a belt of country stretching for a length of 500 km with an average breadth of 150 km from Stavanger in the southwest to Meraker in the Trondhjem district to the northeast. It included the great belt of Caledonian orogeny and igneous action associated with a zone of overthrusting towards the southeast along the Caledonian front. These researches are set down in five monographs under the general title "Geologisch-petrographischen Studien im Hochgebirge des südlichen Norwegens."

In the Finse region, an area of great significance in the elucidation of the tectonics of south central Norway he contrasted the hornfels-like alteration of the Cambrian conglomerates preserved and protected from stress in hollows on the sub-Cambrian land surface with the gneissic deformation of the old granite occurring as disrupted blocks in the schistose alum phyllites of Sandaadalen, above this horizon, and in a later contribution he presented a general picture of the metamorphism of the Caledonian geosyncline. In the Trondhjem district he was able to map zones of progressive metamorphism marked by index minerals of argillaceous sediments, chlorite, biotite, and garnet, unaware, it is clear, of the much earlier zonal work of Barrow in the southeast Highlands of Scotland. The progressive metamorphism of the calc silicate rocks of this same region was exhaustively studied and it is of interest to note that his treatment contains a development of the concept of metamorphic facies proposed by Eskola in 1914 and further elaborated by that investigator in 1920.

In his monograph on the igneous rocks of the Caledonian mountain region his attention was devoted especially to three rock suites he recognized—the "green lavas" and associated intrusives, the Bergen-Jotun series of charnockite-anorthosite type and the opdalite-trondhjemite suite. Their eruptive provinces he showed were partly separated in space and partly overlapped and he recognized a possible genetic relationship between the green rocks and those of the trondhjemite suite. It was a detailed study of the petrography and geological environment of these suites (*Stämme*) which brought him to support the conception of petrographical provinces and to the conviction that differences among igneous rock series were dependent in the first place on geological environment rather than on contrast of original chemistry. In these deductions he was elaborating a thesis earlier developed by Harker.

His trondhjemite suite he recognized as characteristic of the active stages of mountain growth from a geosyncline, and owing its special mineralogical character to the special environment of intrusion among the wet sediments of the geosyncline. On the other hand though its Caledonian age could not be demonstrated, the charnockite-anorthosite suite of the Jotunheim he considered the product of crystallization of a dry magma intruded in an anhydrous environment of older rocks.

An important contribution to the tectonics of the Eastern Jotunheim was made in his study of the *Hoifjeldskvarts* conglomerates. The Bergen-Jotun igneous suite already referred to, has suffered much from tectonic disturbance and in its eastern outcrops the gabbros of the suite lie as sheets over the Palaeozoic sediments and are in places intensely mylonitized on their lower surfaces. Goldschmidt in a study of the region between Gudbrandsdal and Valdres found that the boulders of the conglomerate which overlie Ordovician phyllites consist of a gabbro identical with that seen in the gabbro masses overlying the conglomerate. He was led tentatively to the conclusion that these conglomerates of Caledonian age were formed in front of a thrust gabbro mass impelled from behind and moving at the surface or under cover of water only. The gabbro masses he pictured as occupying a similar tectonic position as many *Klippen* of the Alps and related themselves to the *Hoifjeldskvarts* as the Alpine *Klippen* to the *Flysch,* the gabbro conglomerates being paralleled with the exotic conglomerates. But the Alpine analogy was here exceeded for he recognized the thrust masses of the Eastern Jotunheim as travelling over their own conglomerates. In a neighbouring district he recorded a quartz conglomerate developed in similar relationship to a thrust mass of granite.

Active field work in this geological series of researches was brought to a close in 1917 but the final monograph did not appear till 1920. In his study of the injection metamorphism of the Stavanger region he portrayed convincingly the nature of the metasomatic metamorphism associated with intrusions of the trondhjemitic suite and he set up a new standard of achievement in the investigation of injection complexes by this exhaustive and quantitative study which has served as a model for all later work. The results he demonstrated on the ground, at a memorable international gathering of geologists which he himself had arranged, in the summer of 1922. The realities of metasomatism worked out in this field and in the aureoles of the Oslo igneous massifs led him eventually to present a study on metasomatic processes in silicate rocks, distinguished especially for the recognition and formulation, in the light of the mass action law, of the principle of minimum concentration for effective metasomatism.

Goldschmidt's researches in geochemistry were initiated during the First World War when the deficiency of raw materials led him to extended investigations on the mineral resources of Norway. The organization of this research was entrusted to him by the Norwegian government in 1917 and he became Chairman of the Commission for Raw Materials and Director of the Raw Materials Laboratory. The origins of modern geochemistry may be said to date from this period. After the close of the war he set himself to continue research along more general lines than was possible during the years before—in a study of the general scientific problems intimately connected with the utilization of mineral raw materials. As he himself has written of his programme at that time—"Among the problems proposed there was one of outstanding importance, i.e. to find the general laws and principles which underlie the frequency and distribution of the various chemical elements in nature—the basic problem of geochemistry. I proposed to attack the problem from the viewpoint of atomic physics and atomic chemistry and to find out the relationships between the geochemical distribution of the various elements and the measurable properties of their atoms and ions."

With a small team of co-workers he applied himself with great energy to these studies. To this end both old and new methods of chemical and physical analysis were brought to bear on the investigation of rocks and minerals. X-ray analysis and quantitative spectrographic methods provided new lines of attack. Chemical analysis by X-ray spectra had been first applied to mineral analysis by Hadding working in Siegbahn's laboratory in 1922. Goldschmidt was quick to adopt this new development. His memoir with Thomassen on the distribution of the rare earth metals in minerals, published two years later, demonstrating the greater relative abundance of the elements of even atomic number in this series, in harmony with the rule of Harkins and Oddo, was a singularly successful example of its use. But one of the first problems he attacked at this time was the partition of chemical elements between systems of co-existing liquid phases—molten iron, liquid iron sulphide and fused silicate, a matter of obvious importance in metallurgy, petrology, meteorite study and indeed in planetary chemistry. From experience of metallurgical practice, an intimate knowledge of the composition of meteorites and a mass of geochemical analytical data he was led to subdivide the commoner elements into the groups siderophile, chalcophile and lithophile characteristic respectively of the metal, sulphide and silicate melt. These studies brought a special contribution to the problem of the chemistry of the earth's deep interior for with Tammann he conceived the earth ball as a three phase system, recognizing an intermediate sulphide-oxide shell between the metal core and a silicate outer shell. He now added a further "atmophile" group of elements to his classification, those concentrated in the gas phase of the primordial atmosphere. The principles of partition of the elements among immiscible phases he was later to elucidate more fully and finally to add a fifth "biophile" group of elements to his geochemical classification, those concentrated by living organisms in the later stages of terrestrial evolution.

So were initiated the classical series *Geochemische Verteilúngsgesetze der Elemente* in which he was ultimately to lay the foundation of inorganic crystal chemistry though he recognized it as simply one aspect of the general geochemical problem he had set himself many years before.

Looking back in 1926 he could write "The experimental work in the field of crystal chemistry which in recent years the Mineralogical Institute of the University has undertaken had the objective to throw light on the geochemical distribution and technical properties of practically important materials. During the carrying out of these researches . . . there appeared a series of general crystallochemical relationships which made possible a notable deepening of the principles and data of crystal chemistry."

In the first place it may be imagined that he thought of the X-ray methods as just one of the auxiliary methods for the examination of minerals to be used when other methods failed. He started this work at the beginning of 1923 and from then on until 1929 when he moved to Göttingen, he and his small group of research workers several of whom like T. Barth, Zachariasen and Oftedal were subsequently to become noted X-ray workers, carried out what is still the most remarkable planned series of X-ray studies that has ever been undertaken. Although the work set out in Nos IV to VIII of the "Verteilungsgesetze" seems to show an absolutely straightforward progress it is clear from internal evidence that Goldschmidt enlarged his views as he went along while maintaining the unity and purpose of the project. In the course of the work nearly 200 separate chemical compounds were prepared and analysed by X-rays and the results published in 45 different papers. At the time when Goldschmidt started this work crystal chemistry as a scientific discipline did not exist. As he wrote in his preface to "Verteilungsgesetze" No. VII—"For the mineralogist and for the practical chemist crystallography was a purely descriptive auxiliary science which made possible the recognition and distinction of crystalline materials whether they were minerals or technical products."

He was progressively to give it nearly all its leading ideas. It is characteristic of his geochemical interest that the first series of compounds examined were those of the uranium group of elements—uranium oxide, thorianite, cleveite and bröggerite.* He saw that the similarity between these from the X-ray point of view was much more important than the difference revealed by chemical composition. In the same paper he discusses the close relations between yttrofluorite (YF_3) and fluorite (CaF_2) and thus was immediately plunged into all the difficulties involved with pseudoisomorphism and defect structures. The really effective beginning of the crystal chemical work came, however, with his masterly study of the sesquioxides, first of the rare metals and then of the other trivalent elements. His interest in these was clearly geochemical in the first place, but the magnificent series of rare earth sesquioxides enabled him to show the existence of the three crystal types (*a*), (*b*) and (*c*) in each of which isomorphism went with a steady change in lattice dimensions, the celebrated lanthanide contraction, the significance of which he was the first to point out. It was obvious to him from these extended series in the light of his deep understanding both of geochemical distribution and of the periodic table that both the contraction and the morphotropy of the sesquioxides depended on a continuous change with atomic number in the size of the rare earth atoms. Now atomic size had been discussed even before the first analyses of crystal structure but there had been a very natural confusion between atomic volumes derived from the densities of the elements expressed in the Lother Meyer curve and interatomic distances from binary compounds. This difficulty was not effectively cleared up until the work of Wasastjerna in 1932 [*sic;* 1923(?), ed.] and Goldschmidt was one of the first to make use of Wasastjerna's new picture of large anions and small cations to which he furnished the most extensive series of examples. It was in meditating on the transition from one crystal type to another, brought about by comparatively slight changes in the radii of the cations, that he saw that the laws of co-ordination already postulated years ago by Werner to account for complex salts also held the key to the structure of simple ionic crystals. Very small cations fit between two anions, larger ones between three, four, six and eight. The characteristic mean size for a cation to fit in a definite co-ordination sphere of anions was, he saw, fixed by the simple geometry of the radius ratio of the two ions. Thus all simple binary compounds AX had 3:3 co-ordination if the radius ratio was less than 0.22, 4:4 if it lay between this and 0.41, 6:6 up to 0.73 and 8:8 beyond this. The series did not end there; if, as sometimes is the case, the cation was actually larger than the anion it descended in the same order giving rise to what Goldschmidt was the first to call anti-isomorphous substances which might have the same crystal structure by an interchange of positive and negative ions but which could never, unlike isomorphous substances, give rise to mixed crystals.

Goldschmidt, however, had no sooner found this law than he realized that it held many exceptions and indeed he only made it the starting point of his series of fundamental laws of crystal chemistry. The law broke down under the circumstances which he clearly distinguished—if the anion was large, such as sulphur or iodine, or even more where the anion was a complex such as OH, the anion was deformed by the strong polarizing field of the cation leading to structures which approximated far more to those of the molecules of organic chemistry. This was particularly true where, as in such sheet-like structures as molybdenite, all the cations were on one side of the anions. Goldschmidt was the first to distinguish between the relative importance of the polarizability of the anions which was that already established by Fajans for their optical properties, and the polarizing power of the cation which was directly proportional to its charge and inversely to its radius. In this way he was able to arrive at the general law that "the structure of a crystal is determined through the numerical proportions, the ratio of sizes and the polarizability of its units."

The other region where even this rule breaks down is in those structures where the binding is very markedly non-ionic, particularly in the compounds between the transition elements and those of the IVth, Vth and VIth B groups of the periodic system. Goldschmidt showed that the laws of interatomic distances in such compounds as the compounds of the diamond, wurtzite or nickel arsenide structures obeyed quite different rules from those of ionic compounds. The interatomic distances in fact remained sensibly constant as long as did the ratio of electrons to atoms, whatever the valency of the individual atoms. He was thus led to postulate two incommensurable types of structures, one being the ionic and the other semi-me-

* In his much later published paper "Om super-uraner grunnstoffen med storre Kjerneladning enn 92" he returned to the Uranides and showed how the new artificial elements fitted into the scheme he had elaborated.

tallic or adamantine, though no simple name has yet won general acceptance.

The whole of Goldschmidt's work was summarized in a masterly way in the two great penultimate papers of the "Verteilungsgesetze" VII and VIII. There has rarely been in the history of science an example of an enunciation of fundamental laws combined with a wealth of well-ordered and precise data determined at the same time by the same person. With the publication in 1926 of these two numbers, modern crystal chemistry received its first and definitive form. It contains not only a detailed description of the radii of almost every ion to be found in the periodic table, but of the description and explanation of all the structures for the types AX, AX_2, A_2X_3, ABX_3, ABX_4 and A_2BX_4 which include between them most of the simpler inorganic compounds. It was accompanied by tables and diagrams of atomic radii that became from the start the basic data for all subsequent crystal chemical work. In these two papers Goldschmidt discussed not only the general laws already referred to including those of polymorphy, isomorphism and morphotropy, but showed how they followed from the very nature of the coulomb attractions between the ions, thus doing on an extended scale what Born had achieved with the refinements of mathematical physics for the simpler structures.

In this work Goldschmidt had succeeded in solving the basic geochemical problem that he had put himself at the start of the investigation, namely—what are the characteristics of atoms that determine their distribution in the crust of the earth? He showed that for the most part these were ionic size, charge and polarizability. This work, combined with the knowledge of silicate structures derived from the Bragg school, explained with the utmost elegance and simplicity the puzzling rules of ionic substitution in minerals. The older mineral chemists had used the analogies of organic chemistry in putting forward formulae for the complex minerals such as the micas and feldspars. Goldschmidt showed that they were on the wrong track and in fact a much simpler explanation was that of the piling of spheres in different sizes. The possibility of substituting any one atom by another even of different charge was possible so long as ionic sizes did not diverge too much.

He extended the theory beyond this to explain the physical properties of crystals. He showed, for example, that the hardness of crystal was determined by the charge and distance between the closest ions and established the two parallel series, so called model structures of the relatively soft fluorides with single charges and the hard oxides with double charges, and in which each anion in the first series was matched by another of the same radius and double the charge in the second. Thus the idea of model structures was to prove extremely valuable in the further elucidation of the complex minerals.

Goldschmidt did not stop there, however, he went on in one last but illuminating extensive study of the metallic state, thus rounding off the whole picture of inorganic chemistry. Here, guided by this idea of the importance of co-ordination he was able to establish sets of atomic radii corresponding not now to different atomic charge but to different co-ordination number and thus provide the most beautiful illustration of the regularities of the periodic system of the elements, free from anomalies produced by accidents of different crystal structures. Thus bismuth, which normally crystallizes in an irregular 6 co-ordination can be made to alloy in 12 co-ordination with a reduction in effective volume of 8%. Having thus put light and order into a field of study which hardly existed before his time, Goldschmidt after 1928 returned to his main preoccupation with geochemical problems. His work, however, was at once seized on by crystallographers and crystal chemists and made the basis for all subsequent studies. Neither in general principle, nor in any important particular, has it been shaken to this day. The first great work of Pauling was in extending the conceptions of Goldschmidt to cover the more complex silicates whose structures had been determined by the Bragg school. Naturally this involved refinement of the purely empirical atomic and ionic radii determined by Goldschmidt but these are still the most convenient for practical use. His work on the semi-metallic and metallic structures has also stood the test of time and many of his minor interests such as those of non-Daltonian ratios in minerals have proved most fertile. Goldschmidt's work, however, in this field is much more than the establishment of the fundamental rules of crystal chemistry and the determination of its constants for nearly all the elements. He built the foundation on which modern inorganic chemistry rests and we may soon hope to see his ideas taken as the basis for the elementary as much as for the advanced teaching of inorganic chemistry.

In the third main period of his scientific activities

Goldschmidt worked in Göttingen. On the retirement of Mügge in 1928 he was invited to fill the Göttingen chair and he became director of a large new mineralogical institute which he organized on the model of his Oslo laboratory.

He moved there in 1929 and began a long series of geochemical investigations on the rarer elements. For the most part his methods of analysis were now founded on carbon arc spectrography. With his colleagues Mannkopff and Peters he developed the cathode layer technique, capable of detecting most elements in concentrations down to 0.001% or even less and adapted for the treatment of much smaller quantities than hitherto in spectrographic work. A stream of publications dealing with the geochemistry of selected elements, among them, gallium, germanium, scandium, beryllium, boron, the noble and the rarer alkali metals, was issued during the years 1930–1935 in the Göttingen *Nachrichten.* It is not too much to say that the elaborate series of investigations mainly carried out at this time has revolutionized our knowledge of the distribution of the minor constituents of the earth's crust. They were based on a wide selection both of terrestrial and meteoritic matter and he spared no pains in gathering materials for research from all quarters of the globe. He was ultimately to present a broad synthesis in a general survey tracing the partition of a large number of these rarer elements among the successive products of the geological cycle from the primitive magma to the sediments and the sea. He drew an illuminating picture of the sifting and sorting of the elements in the fractional crystallization of igneous magma and demonstrated the effects of ionic size and charge in regulating the sequence of entry of elements in the host silicate structures of the consolidating igneous rock. He essayed a quantitative balance of substances through a succeeding stage of weathering and sedimentation—in the residual sediments and through solutions to the sea—and employed the function of ionic potential, the quotient of ionic charge and size, introduced into physical chemistry by Cartledge, as a guiding principle in the distribution of the elements at this "aqueous" stage of geochemical evolution.

It was on the basis of the ionic potential that he drew distinction between three major groups of elements in the process of sediment formation, those remaining in true ionic solution, those precipitated by hydrolysis and those forming soluble anionic complexes, groups which can be clearly recognized as common products in distinctive units of the sedimentary record. The principle thus empirically derived, and in his hands successfully exploited, has yet to be grounded with an adequate theoretical foundation, but the effective use he made of his own fundamental discoveries in crystal chemistry in interpreting his data and indeed in predicting geochemical phenomena, is apparent on every hand in these comprehensive researches.

He went on in sequence to the consideration of a further stage of the geochemical evolution of matter in his analysis of the processes of concentration of elements due to the action of living organisms—the province of biogeochemistry. His interest here had been stimulated by his discovery that germanium occurs concentrated in certain coal ashes—such as those of Northumberland and the Donetz basin. He found that a concentration of a great number of minor elements occurred in forest humus and forest litter and traced a circulation of these elements in solution from the subsoil through the plant to ultimate precipitation and enrichment in the topmost layers of forest soils. The phenomena of enrichment thus disclosed have in recent years been successfully adapted in a type of geochemical (spectrographic) prospecting where a systematic search for a concentration of trace elements in vegetation or soil humus may give clue to the site of an underlying and hidden mineral deposit.

Goldschmidt's term at Göttingen came to an end in 1935. In the face of growing Nazi anti-Semitism he resigned his chair and returned to Norway, regaining Norwegian citizenship. A little later (1936) he succeeded to a professorship which became vacant at Oslo on the death of Schetelig. He set up a spectrographic unit as a section of the Norwegian government's economic laboratory and continued research along the lines established at Göttingen. Working with renewed energy he brought out the ninth and final publication in the "Verteilungsgesetze" series. In this work he linked up a mass of geochemical and astrophysical data as well as data from nuclear physics and he attempted an estimate of the universal mass ratios not only between the chemical elements but also of the isotopes, devoting some attention finally to the interrelation of the frequency of the different isotopes in the universe, their stability and the conditions of their formation.

During the period of his association with the government laboratory in Oslo he did much to stimulate interest in, and to develop, the mineral resources of Norway. He was a pioneer in the development of forsterite as a commercial refractory and spent much time on the problem of the utilization of the large Norwegian deposits of olivine on the west coast, particularly those of Sondmøre. Already in 1928 he had made industrial experiments on the use of olivine sand and crushed dunite in place of quartz for making moulds for foundries but these results had remained unpublished. In the succeeding years the industrial use of olivine refractories was largely developed in Germany, utilizing his methods of transforming olivine rock into a mixture of forsterite and magnesium ferrite by burning with magnesia. Such refractories have had successful use in ceramic kilns and for the roofs of copper-refining furnaces among other adaptations. One of his last papers (1945) issued in collaboration traced the effect of olivine on the lungs of rats, a test of olivine dust in problems of silicosis which showed that its substitution for quartz in some foundry processes should be effective in reducing the incidence of that disease.

He worked on the development of titanium pigments, the importance of mica minerals as a source of potassium for plants and suggested the possibility of using biotite to this end. Though his suggestion was not altogether new the resulting experiments he developed are probably the most complete set yet carried out to test the possibilities. Among other developments were his process of making alumina from anorthosite, a rock in which Norway was abundantly supplied.

During the Second World War and after the German occupation of Norway, Goldschmidt suffered severely at the hands of the Nazis. He was twice arrested and finally in October 1942 was sent to a concentration camp, the whole of his property being confiscated. At the end of November of that year he was sentenced to immediate deportation to Poland. At the last moment this order was cancelled and he succeeded just before Christmas in escaping across the frontier to Sweden. He came to Britain in the spring of 1943 and offered his services to the Allies. Here he worked under the auspices of the Agricultural Research Council, spending the first year at the Macaulay Institute for Soil Research, Aberdeen, and later went to Rothamsted Experimental Station where he remained until his return to Oslo in June 1946. His experiences during the war years had led to a serious deterioration in his health and when he returned to Norway he was very weak after long and serious periods of illness. He seemed to be making a good recovery after an operation and his sudden death on 20 March 1947 came as a shock to his friends who saw him regularly.

For some years before his death Goldschmidt had been engaged on a comprehensive treatise on geochemistry, synthesizing his researches of 25 years. The greater part of it was in draft before he left England and more was completed in his last days in Norway. These chapters are now being translated or corrected for the press and it is confidently expected that they will soon be published in this country.

Goldschmidt was a man of amazing energy and fertility of ideas. Not even periods of illness could diminish the ardour of his mind, incessantly directed to the solution of problems he set himself. First and last a scientist, he viewed everything in relation to science and research. A complete list of his publications runs to more than 200 titles, many of them covering monographs of great length. Within his own fields his knowledge was phenomenal and his memory stupendous and he gave the impression that he had most of the existing literature in his mind. Professor Born has related that at Göttingen he once sent an assistant to Goldschmidt to ask about literature on a great number of data. Instead of giving the titles of papers Goldschmidt just dictated numerical tables from memory and a later check proved them essentially correct. He knew by heart not only the properties of atoms and ions, like radii and polarizabilities, but also their intimate occurrence in minerals, and it seemed that if he did not know a fact he had some simple way of deriving it from his radii. He remembered at once where he had buried a paper he wanted, and this was all the more astonishing as he had a system not to tidy up a writing-desk but to start a new one, when the old one was piled high with papers. So gradually nearly every room in his house came to have a writing-desk until there was only a kitchen sink in an unused kitchen left and even this was covered with a board and turned to the prescribed use.

In conversation one might get the impression that in his enthusiasm his ideas were running away with him but such exuberances were never committed to print where he was most careful in weighing his words. Scientific honesty was indeed with him a first principle.

At home in Göttingen he lived with his old father, as if they were two brothers—in fact they came to be known as Old-Brother and Young-Brother. In the University Faculty meetings he was always quiet and polite; he spoke seldom, and if he spoke, he did so slowly, with great deliberation and clarity.

His addresses given to the Gesellschaft der Wissenschaften were brilliant, a wonderful mixture of exact measurements and the keenest combinations of ideas to interpret them. He often said that he was happy in Göttingen because he had in the Gesellschaft a group of men who understood and appreciated his work as indeed they did.

He had a great fund of humour which took at times an extravagant or caustic form brought out in some of the legion of Goldschmidt stories. It is reported that when Hitler first came into power Goldschmidt appeared at a Göttingen meeting with his hair cut more than usually short. His colleagues commenting upon it, he tapped his head and said, "Damit es besser rollen kann"; and after, when he was in England and returning to Norway, he remarked, "I owe two things to Hitler; first, without him, I would be a geheimrat professor with nothing to look forward to but ease and honour. Now I have lost everything and have the chance to be young again," and he added, "Besides it has taught me a lot; you see, in ordinary life there is every inducement for people to behave decently, but under Nazi-ism it is quite the contrary and yet I find a number of people who do behave decently—even among the educated classes." His love of, and patient approach to, animals was great. He had tamed birds and squirrels in his garden and these were all given names which had some funny allusion to people he knew. If, on occasions, in his judgment of men and situations he was apt to exaggerate and react childishly, it is to be remembered that the lapses were those of a generous but sensitive nature, and that he was quick to acknowledge his mistakes and make amends. To those who came to understand this, he was a devoted and loyal friend.

He ruled in his department with benevolent tyranny and his pupils and collaborators adored him. He took a keen interest in their welfare and they became his personal friends. In his recognition of their work he was liberal and was ever ready to place his knowledge at their disposal even though at the time deeply engrossed in other problems.

When he came to Britain his health was already seriously undermined. He deeply appreciated the help extended to him at the Macaulay Institute and at Rothamsted, and while his main preoccupation was the completion of his *Geochemistry* he spent much time with his colleagues, was generous with his advice, and took great interest in their work of trace element distribution.

Honours came to Goldschmidt from many countries, among them honorary or foreign membership of most of the leading scientific academies of Europe. It is pleasant to remember that in this country his great contributions had been recognized by his electron as a Foreign Member of the Royal Society in 1943 and of the Geological Society of London from whom he received the Wollaston Medal, its highest award. He was invited to deliver the Hugo Müller lecture before the Chemical Society in 1937 and was the recipient of an honorary degree from Aberdeen University. In his own country he had received the Nansen Prize and was made a Knight of the Royal Norwegian Order of St Olaf.

For the appreciation of Goldschmidt's work in the field of crystal chemistry, the writer is indebted to Professor J. D. Bernal, F.R.S., and he owes acknowledgment to Dr I. Oftedal of Oslo for important data on Goldschmidt's life and work in Norway, to Professor M. Born, F.R.S., for information on his period at Göttingen, and to Dr W. G. Ogg, on his period at Rothamsted.

Commemorative speech on Professor V. M. Goldschmidt at the Mathematics and Science Conference of 14 November 1947*

O. HASSEL

WHEN Victor Moritz Goldschmidt died on March 20 this year (1947), our Academy lost a well-known member, a natural scientist who left a deep imprint on the world of science.

For many years he had instilled in himself a great desire for work, a greater desire than was tenable and, in addition, the strain of the war proved too much for him. When he returned to Norway, in the summer of 1946, he was already a dying man. We felt that, in a way, an important part of science was buried when we realized he was no longer with us. Goldschmidt had represented science in a way true to our best tradition. It is not easy to express the loss our scientific life has suffered by his death at only 59 years of age.

Goldschmidt was not Norwegian by birth, and even though he didn't feel himself to be truly Norwegian, in his work as a geologist and mineralogist he was indisputably so. The foundation built by Th. Kjerulf, W. C. Brøgger and J. H. L. Vogt ensured that Norwegian science was highly respected in these fields. This, along with the nature of the country, ensured that it was not his father's field of science, chemistry, that originally drew him to begin university in Oslo in 1905 at the age of seventeen. But it was chemistry, especially inorganic and physical chemistry, that soon began to interest him strongly, and his insight and knowledge in these fields was impressive. In his later years after his geochemical work had brought him in close contact with the problems of nuclear physics, he threw himself energetically into the study of quantum mechanics.

Goldschmidt's scientific efforts span so many fields that it is unlikely anyone could be found among us who was well acquainted with all of it. His first major work, his thesis, was printed in the Academy's journal in 1911. This work, which dealt with contact metamorphism in the Oslo area, and as has been previously stated (by others), is something out of the ordinary. Although it is based on a great deal of material, each aspect is, nevertheless, carefully documented from both a mineralogical and chemical point of view.

His great petrographic conclusions are based on the principles of physical chemistry and apply far beyond the special conditions of the Oslo area. Goldschmidt's work on contact metamorphism expanded, from 1911 to 1921, into studies which include all the mountains in the south of Norway, as far as the Trondheim area, and contain thorough tests of regional metamorphism. In these works, he introduced the idea of the origins of certain species of rocks which form from genetic and chemical-mineralogical related igneous rocks, and he showed that at least three such types could be identified in the mountains of southern Norway. These works made a considerable impact, especially his studies of the so-called "injection-metamorphism" in the Stavanger region. After Goldschmidt had become the Chairman of the State Raw Materials Committee in the fall of 1917, he concentrated more and more on problems connected with the occurrence of raw materials in the crust of the Earth. The comprehensive scientific investigations he initiated in this regard resulted in important knowledge of the geochemical distribution of the elements. With a few fellow workers, these geochemical projects were continued under his knowledgeable leadership for the next 25 years up until his death. For the first twelve years, until Goldschmidt moved to Göttingen in 1929, he carried out his work at the University's Mineralogical Institute, making it the real home of geochemistry. At this time, X-ray methods were just beginning to be used, for example, to test spectrographic and crystal-structure data. From very generalized data based on extremely precise experimental methods, Goldschmidt succeeded in contributing fundamental ideas to geochemistry and to the theory of the relation between crystal structure and chemical characteristics. It is accurate to say

* From: *Det Norske Videnskaps-Akademi i Oslo,* Årbok 1947, pp 85–102. Translated by Olaf Niemenn and Bernard Trevor, Alberta Research Council.

that he, more than anyone else, is the creator of crystal chemistry.

Goldschmidt divided the elements into siderophile, chalcophile, lithophile and atmophile, and he succeeded in finding the relations between these divisions and each element's electronic characteristics. Probably the most interesting parts of his works are those which deal with the nature of geochemistry and crystal chemistry in various parts of the Earth. Especially important here is the reduction in the ionic radius of an ion in the lanthanide series (La-Lu) which occurs as the atomic numbers increase. This "lanthanide-contraction" is of great importance in the relations between elements. The knowledge of atomic and ionic radii was of great importance in all later work on crystal structures. Goldschmidt was quick to realize the importance of radius ratios for coordination numbers, and he explained why certain limited values responded to the changes between certain structure types. By also taking into consideration the polarization qualities of single ions it became possible to explain the relation between the configuration of the single atom, its ion and its chemical characteristics. On the basis of this, he improved on theories of isomorphism, polymorphism and morphotropism, and the relations between the physical characteristics of a crystal and its chemical composition.

After Goldschmidt took over the professorship in Göttingen, he set up an institute like the one in Oslo but including, as well, a large and well-equipped department for optical spectrographic analysis. In the years leading up to 1935 he carried out a great deal of specific quantitative analyses on the abundance of the elements in the crust of the Earth, and through the analysis of meteorites, of material from outside the Earth. Important advances in quantitative spectrographic analysis methods were also made during this period.

After his return to Norway in 1935, he published No. 9 of his "Geochemische Verteilungsgesetze der Elemente," printed in the Academy's journal. In this work he used all available data which he considered reliable, whether his own or that of others. He also was concerned with astrophysical, and the newer nuclear-physical data. He tried in one paper to reach as broad and reliable a view as possible on the connections among all the chemical isotopes in the universe. The value of these geochemical studies depended as much upon the reliability of his own experimental contributions as on his critical ability to judge the precision and range of the results of others. His broadly-based knowledge was extremely important here.

Although Goldschmidt's practical work has not become as important as his scientific research, its influence is nevertheless stimulating. In one field he saw for himself the practical results of his work: the use of the Norwegian "olivinsten" (olivine stone) to make fire-resistant materials.

A unique abundance of ideas and daring characterized Goldschmidt's attack on difficult problems. His scientific enthusiasm was, however, controlled by his intelligence and critical judgement. He also had an extraordinary memory.

It is seldom that a researcher can accomplish in one life work which is so versatile and broad, and still possess an inner connection as organic as Goldschmidt's. And he completed all this before he was fifty years old. Then the war came with all its accompanying stress. It is tragic that the events of these later years prevented him from completing the more generalized comments which we know he was going to write. An important part of science has therefore been lost.

The critical common sense which was characteristic of Goldschmidt as a scientist was not always apparent in his judgement of situations and people. This aspect of his person showed a lack of understanding. Often he admired too easily and, even more often, he was too suspicious of the good intentions of others for no apparent reason. He was especially suspicious of one of his co-workers from Göttingen, who was aware of this. The latter wrote in a letter shortly after Goldschmidt's death "But in his heart he was decent." It is certain that many of us have reason to thank him for the unselfish help he gave when he was asked a scientific question.

The position he achieved in scientific circles abroad is unique. One way to observe this is through all the awards he received. But also, on a human level, he made many good friendships with fellow workers from abroad. Norwegians who visit other countries can attest to this.

We honor his memory.

Appendix B

Bibliography of Victor Moritz Goldschmidt Compiled by W. P. Sommerfeldt (from *Norsk Geologisk Tidsskrift,* 1949, 27, 150–163)

1906

Die Pyrolumineszenz des Quarzes, 19 s. *Forh. VidenskSelsk Krist.* 1906. No. 5.

1907

Über Albit von Grönland, Af V. M. G. og C. Dreyer., *Medd. Grønland,* 1907, **34,** 1-60.

1908

Profilet Ringsaker–Brøttum ved Mjøsen. (Med et geologisk kart. Resumé in deutscher Sprache.) *Norg. geol. Under.,* 1908, **49.** Kra. 1909. II, 11–40.

Radioactivität als Hilfsmittel bei mineralogischen Untersuchungen. Hierzu 1 Textfigur. I. II. *Z. Krystallogr. Miner.,* 1908, **44,** 545-560, Tai. X, **45,** 490–494.

Über Argyrodit aus Bolivia. Mit 1 Textfigur. *Z. Krystallogr. Miner.,* 1908, **45,** 548-554.

1909

Geologiske iagttagelser fra Tonsaasen i Valdres. Med figurer i teksten og en geologisk kartskisse. Resumé in deutscher Sprache. *Norg. geol. Unders.,* 1909, **53,** Kra. 1910. III.

Über Mineralien der Kontaktmetamorphose im Kristiania–Gebiet. (Vorläufige Mitteilung.) *Zentbl. Miner. Geol. Paläont.,* 1909, **13,** 405–410.

1910

Notis om kvalitativ bestemmelse av kiselsyre. *Tidsskr. kemi Farm. Terapi,* 1910, **7,** 325–326.

Über den Unterschied von Storaxzimtsäure und synthetischer Zimtsäure. Von Rüber & V. M. G. *Ber. chem. Ges.* Berlin, 1910, 453–462.

1911

Die Kontaktmetamorphose im Kristianiagebiet. Mit 84 Abbildungen im Text. 2 Tafeln und 5 farbigen Karten. IX + 483 s. 5 kart, pl. I–II. Doktoravhandling. *Vidensk. skrifter I. Mat.-naturv. klasse,* 1911. No. 1.

Petrographische Untersuchung einiger Eruptivgesteine von Nordwestspitzbergen. Mit einer Tafel. Utgitt for Fridtjof Nansens fond. 17 + upg. s. pl. *Vidensk. skrifter I. Mat.-naturv. klasse,* 1911. No. 9.

Professor dr. W. C. Brøgger. *Tidssk. kemi Farm. Terapi,* 1911, **8,** 337–340.

Über isomorphe Mischungen von Epidot und Orthit. *Zentbl. Miner. Geol. Paläont.,* 1911, **1,** 4–5.

Über den sogenannten Sphenoklas. *Zentbl. Miner. Geol. Paläont.,* 1911, **2,** 35–36.

Die Gesetze der Mineralassosiation vom Standpunkt der Phasenregel. *Z. anorg. Chem.,* 1911, **71,** 313–322.

Anwendung der Phasenregel auf Silikatgesteine. *Z. Elektrochem.,* 1911, **17,** 686–688.

1912

Geologisch-petrographische Studien im Hochgebirge des Südlichen Norwegens. Utgitt for Fridtjof Nansens fond. I. Ein kambrisches Konglomerat von Finse und dessen Metamorphose. Mit 5 Taf. 18 s. 5 pl. II. Die kaledonische Deformation der südnorwegischen Urgebirgstafel. *Vidensk. skrifter. I. Mat.-naturv. klasse,* 1912. No. 18, 19.

Die Gesetze der Gesteinsmetamorphose mit Beispielen aus der Geologie des südlichen Norwegens. (Mit einer Textfigur.) Utgitt for Fridtjof Nansens fond. *Vidensk. skrifter. I. Mat.-naturv. klasse,* 1912. No. 22.

Über krystallographische Untersuchungen bei tiefen Temperaturen. 6 s. *Forh. Vidensk. Selsk Krist.,* 1912. No. 1.

Om katalytiske virkninger. Foredrag 26. jan. 1912. *Overs. Vidensk. møter i* 1912. Kra. 1913, 1–2.

Lovene for bergarters metamorfose i jordskorpen og deres anvendelse paa det sydlige Norges geologi. Foredrag 22. nov. 1912. *Overs. Vidensk. møter i* 1912. Kra. 1913, 67.

Über die Anwendung der Phasenregel auf die Gesetze der Mineralassoziation. *Zentbl. Miner.*, 1912, 574–576.

Über die Winkeländerung der Krystalle bei tiefen Temperaturen. *Z. Krystallogr.*, 1912, **51**, 1–27.

Über ein mehrkreissiges Goniometer und seine Anwendung. *Z. Kristallogr.*, 1912, **51**, 28–39.

Über Quarz von Finse in Norwegen. *Z. Kristallogr.*, 1912, **51**, 40–48.

Erwiderung. *Z. Kristallogr.*, 1912, 51.

1913

Das Devongebiet am Röragen bei Röros. Mit einem paläobotanischen Beitrag: Die Pflanzenreste der Röragen-Ablagerung von A. G. Nathorst. Mit 3 Textfiguren, 5 Tafeln und 2 geologischen Karten. Utgitt for Fridtjof Nansens fond. 27 + upg. s. pl. I–V, 2 karter. *Vidensk. skrifter. I. Mat.-naturv. klasse,* 1913. No. 9.

Om et devonfelt ved Røragen nær Røros. *Norsk geol. Tidsskr.* **2.** No. 13, 33–35.

Om devonformationen ved Røros. Foredrag 7. nov. 1913. *Overs. Vidensk. møter i 1913.* Kra. 1914, 73.

Zu Herrn Joh. Königsberger's geolog. Mitteilungen über Norwegen. Von V. M. G., J. Rekstad, Th. Vogt. *Zentbl. Miner.*, 1913, 324–328.

Zu Herrn P. Nigglis Abhandlung: Über metamorphe Gesteinsserien. *Tschermacks Mitt.*, 1913, **31,** 695.

1914

Om mineralogiens opgaver. Utdrag av tiltredelsesforelesning holdt 28. sept. 1914. *Naturen,* Bergen, 321–334.—*Bergverksnyt* **7**, 198–199, **8,** 5–7.

Nochmals Hernn Königsberger's geologische Mitteilungen über Norwegen. Von J. Rekstad, Th. Vogt. *Zentbl. Miner.*, 1914. No. 4, 114–118.

Om Devonformationen vid Röros i Norge. Foredrag. *Geol. För. Stockh. Förh.*, 1914, **36,** 17–20; 20–23: diskusjon.

Über einen Fall von Natronzufuhr bei Kontaktmetamorphose. *Neues Jb. Miner.*, 1914, **39.** Beilage. (Festbd. Bauer.) Stuttgart, 193–224.

1915

Über die Struktur der Krystalle. Mit 2 Textfig. 27 + upg. s. *Forh. VidenskSelsk. Krist.*, 1915. No. 3.

Om krystallernes indre struktur. Foredrag 29. jan. 1915. *Overs. Vidensk. møter i 1915.* Kra. 1916, 5–6.

Über die Krystallform der Phenylglyzerinsäure $C_6H_5(OH)HC(CH.)HCOOH$ (Smp. 141° C) und ihrer aktiven Komponenten. *Z. Kristallogr.*, 1915, **55,** 123–131.

1916

Geologisch-petrographische Studien im Hochgebirge des südlichen Norwegens. III. Die Kalksilikatglimmerschifer des Trondhjem—Gebiets. Mit zwei Taf. 1916. 2 bl. 32 s. Tabel. Taf. I–II. *Vidensk. skrifter. I. Mat.-naturv. klasse,* 1915. No. 10.

Geologisch-petrographische Studien im Hochgebirge des südlichen Norwegens. IV. Übersicht der Eruptivgesteine im kaledonischen Gebirge zwischen Stavanger und Trondhjem. Mit 2 Fig. im Text, 6 Tafeln und einer geolog. Übersichtskarte. Utgitt for Fridtjof Nansens fond. Bl. 140 s. Pl. I–VI. kart. bl. *Vidensk. skrifter. I. Mat.-naturv. klasse,* 1916. No. 2.

Konglomeraterne inden høifjeldskvartsen. 61 s. pl. I–VII. Kart. *Norg. geol. Unders.*, 1916, **77.**

Om eruptivbergarternes naturlige system. *Forh. skand. Naturf. 16. møte i Krist. 1916.* Kra. 1918, 481–489.

Fører ved geologisk ekskursjon til Finse. Bilag 2 til *Forh. skand. Naturf. 16 møte i Krist. 1916.* Kra. 1916, 1–10, 3 pl.

Fører ved geologisk ekskursjon til Finse. *Nyt Mag. naturvid.*, 1916, **53,** 109–118, 2 pl.

1917

Om høifjeldskvartsen. I og II. Foredrag. *Norsk. geol. Tidsskr.* (1916/17) **4,** 44–46, 49–53.

1918

Stavanger. Geol. rekt.-kart. 6 D, 1:100 000. *Norges geol. unders.*

Nye mineralske raastoffer i Norge. Foredrag. *Beretning om utstillingen for industriel selvhjælp* 4–16 april 1918 *med de i forbindelse med utstillingen avholdte foredrag. Utgit av Kra. Haandverks- og Industriforening.* Kra. 1919, 168–181.

Nye mineralske raastoffe i Norge. Foredrag i Norsk kemisk selskap. *Bergverksnyt,* 1918, 73–75.

Mellem to tidsaldre. Foredrag 3. mai 1918 i Videnskaps-Akademiet. *Samtiden,* 1918, **29,** 405–417.

Teknisk-videnskabelig forskningsarbeide i utlandet og i Norge. *Samtiden,* 1918, 619–629.

Trondheimsfeltets geologi. Diskusjon i Norsk geologisk forening 9. mars 1918. *Norsk geol. Tidsskr.*, 1918, **5,** s. 393.

Om de mekaniske principer for skred og utrasninger. Ref. av foredrag i Norsk geologisk forening 30. nov. 1918. *Norsk geol. Tidsskr.*, 1918, **3,** 401–403, disk.: 403–406.—*Naturen* **43,** 248–253.

Et hittil ukjendt omraade av kalibergarter. *Overs. Vidensk. møter i* 1918. Kra. 1919, s. 20.

En foreløbig meddelelse om Fensfeltet. Foredrag 6. des. 1918. *Overs. Vidensk. møter i 1918.* Kra. 1919, s. 30.

Om Raastoffkomitéens arbeide og resultater. Foredrag i Polyteknisk forening 29. okt. 1918. *Tidsskr. Kemi, Farm. Terapi,* 1918, **15,** 325–339.—*Tek. Ukebl.* **37,** 6–10.

1919

Om aluminiumfremstilling av norske raastoffer. Foredrag i N. I. F., Kra. avd., 24. jan. 1919. *Tidsskr. Kemi,* 1919, **16,** 21–34.—*Tek. Ukebl.*, 1919, **37,** 69–72.—*Materialforvalteren* **5,** 12–14.—*Bergverksnytt* **12,** 3.

Nye mineralske raastoffe i Norge. Foredrag ved Norsk kemisk Selskabs 25-årsjubileum 2. mai 1918. *Tidsskr. Kemi,* 1919, **16,** 65–75.

De hvite farver i naturen og tekniken. Foredrag i P. F. 28. okt. 1919. *Tidsskr. Kemi,* 1919, **16,** 269–277, 281–289.

Stavangertraktens geologi og bergarter. Ref. av foredrag 30 nov. 1920. *Forh. VidenskSelsk. Krist.* 1920. Kra. 1921, 33–34.

Om reaktionen mellem svoveljern og kulsyre. *Tidsskr. Kemi* **17,** 215–218, **18,** 221–228.

Forslag til skandinavisk samarbeide på den prekambriske geokronologiens område. Ref. af foredrag i Geol. fören. 8 april 1920. *Geol. För. Stockh. Förh.,* 1920, **42,** 189–190.

1921

Geologisch-petrographische Studien im Hochgebirge des südlichen Norwegens. V. Die Injektionsmetamorphose im Stavanger-Gebiete. Mit 15 Taf. und einer geologischen Karte. Utg. for Fridtjof Nansens fond. 1921. 142 s., pl. I–XV. Kart. *Vidensk. skrifter. I. Mat.-naturv. klasse.* 1920. No. 10.

Die weissen Farben in Natur og Technik. Leipzig 1921, 20 s.

Om Stavangertraktens geologi og bergarter. Ref. av foredrag i Norsk geol. foren. 11. des. 1920. *Norsk geol. Tidsskr.,* 1921, **6,** 261–263.

Trondheimsfeltets geologi. Diskusjon i Norsk geol. foren. 12 nov. 1921. *Norsk geol. Tidsskr.,* 1921, **6,** 280–281.

Fosfatraastoffene. *Tidsskr. Kemi Bergv.,* 1921, **1,** 4–9.—*Tek. Ukebl.,* 1921, **39,** 119–121.

Om metasomatosen i silikatbergarter (Metasomatosis in silicatbearing rocks). *Geol. För. Stockh. Förh.,* 1921, **43** (1921), 463–468.

Die Phosphatrohstoffe. *Naturwissenschaften.* Berlin. Bd. 9. 6 s.

Krystallographie und Metallkunde. *Z. Metallk.* 7 s.

Ein Reisemikroskop für Geologen und Bergingenieure. *Z. prakt. Geol. Halle (Saale),* 1921, **29,** 187–190.

1922

Stammestypen der Eruptivgesteine. Mit 4 Textfig. Utg. for Fridtjof Nansens fond. 12 s. *Vidensk. skrifter. I. Mat.-naturv. klasse.* 1922. No. 10.

Der Stoffwechsel der Erde. Mit 2 Textfig. Utgit for Fridtjof Nansens fond. 25 + upg. s. *Vidensk. skrifter. I. Mat.-naturv. klasse.* 1922. No. 11.

Om fremstillingen av bariumlegeringer. En tekstfig. og "Zusammenfassung". 34 s. *Norges geol. Unders. No. 107. Statens raastofkomité. Publikation 7.*

Glimmermineralernes betydning som kalikilde for planterne. Av V. M. G. og E. Johnson. Med 1 tekstfig. og Zusammenfassung. 89 s. *Norges geol. Unders. No. 108.—Statens raastofkomité. Publikation 8.*

Das Vorkommen des Elements No. 72 (Hafnium) im Malakon und Alvit. Von V. M. G. und L. Thomassen. *Norsk geol. Tidsskr.* **7** (1922 og 1923), 61–68.

Nordlandsfjellkjedens statigrafi og tektonikk. Diskusjon i Norsk geol. forening 2. febr. 1922. *Norsk geol. Tidsskr.,* 1922, **7,** 72.

Til Regjeringen og Stortinget (om å søke kullfeltene på Svalbard og Bjørnøya bevart for vårt land. Undertegnet av Norsk geologisk forenings styre og tilkalte medlemmer). *Norsk geol. Tidsskr.* **7,** 85–86.

Oversigtskart over utbredelsen av de forskjellige kalimineraler i norsk fjeldgrund. *Norsk Landmandsblad,* 1922, **41,** 268–269.

Hvorfra faar plantene kali? Foredrag 20 jan. 1922. *Overs. Vidensk. møter i 1922.* Kra. 1923, 8–10.

Hvorfra faar planterne sit kali? *Tidsskrift for det norske landbruk* **29,** 97–104.—*Tidsskrift for Skogbruk* **30,** 97–105.—*Norsk Landmandsblad* **41,** 254–256, 267–268.

En sammenlignende undersøkelse over jordklodens bygning og meteoritenes struktur. Foredrag 8. des. 1922. *Overs. Vidensk. møter i 1922.* Kra. 1923, 40–41.

On the metasomatic processes in silicate rocks. *Econ. Geol.,* 1922, **7,** 105–123.

Über die Massenverteilung im Erdinneren, vergleicht mit der Struktur gewisser Meteoriten. *Die Naturwissenschaften,* 1922, **10,** 918–920.

Der Stoffwechsel der Erde. *Z. phys. Chem.,* 1922, **55,** 582–583; *Z. Electrochem. angew. phys. Chemie,* 1922, **28,** 411–421.

1923

Die Krystallstruktur natürlicher und synthetischer Oxyde von Uran, Thorium and Cerium. Von V. M. G. and L. Thomassen. Mit 1 Taf. 48 s. Tab 1. *Vidensk. skrifter. I. Mat.-naturv. klasse.* 1923. No. 2.

Geochemische Verteilungsgesetze der Elemente (I). 17 + upg. s. *Vidensk. skrifter. I. Mat.-naturv. klasse.* 1923. No. 3.

Om nogen dioksyders krystalstruktur. Foredrag 19. jan. 1923. *Over. Vidensk. møter i 1923.* Kra. 1924, s. 5.

Om forekomsten av grundstoffet No. 72, Hafnium, i norske mineraler. Foredrag 16. febr. 1923. *Overs. Vidensk. møter i 1923.* Kra. 1924, s. 7.

Geokemiske fordelingslover. Foredrag 25. mai 1923. *Overs. Vidensk. møter i 1923.* Kra. 1924, s. 25.

Glimmermineralens betydning som kalikilde för vegetationen. Ref. av foredrag. *Geol. För. Stockh. Förh.,* 1932, **45,** 387–389.—*Nordisk Jordbrugsforskn.* Kbn., 382–386.

1924

Geochemische Verteilungsgesetze der Elemente. 2. Beziehungen zwischen den geochemischen Verteilungsgesetzen und dem Bau der Atome. Mit einer Tafel. Utgit for Fridtjof Nansens fond. 37 + upg. s. Tafel 1. *Vidensk skrifter. I. Mat.-naturv. klasse.* 1924. No. 4.

Geochemische Verteilungsgesetze der Elemente. 3. Röntgenspektrographische Untersuchungen über die Verteilung der seltenen Erdmetalle in Mineralen. Von V. M. G. und Lars Thomassen. (Mit einer Textfig.) Utg. for Fridtjof Nansens fond. 58 s. *Vidensk. skrifter. I. Mat.-naturv. klasse.* 1924. No. 5.

(Innlegg i diskusjonen etter Th. Vogts foredrag 2. febr. 1922 Om Nordlandsfjellkjedens stratigrafi og tektonic.) *Norsk geol. Tidsskr.*, 1924, **7**, 72.

(Innlegg i diskusjonen etter Olaf N. Roves foredrag 4. des. 1924: Om de sedimentpetrografiske metoder og deres anvendelse på norsk leir.) *Norsk geol. Tidsskr.*, 1924, **8** (årg. 1924 og 1925), 135.

Om lerjordmineralers opløslighet i syrer og om fremstilling av lerjord av lere og andre raastoffer. *Tidsskr. Kemi Bergv.*, 1924, **4**, 9–11.—*Tek. Ukebl.* 95–96.

Om lerjordmineralers opløslighet i syrer. *Tidsskr. Kemi Bergv.*, 1924, **4**, 9–11.

Om fremstilling av lerjord av lere og andre raastoffer. *Tiddskr. Kemi Bergv.*, 1924, **4**, 38.

Aluminiumfremstilling av lere og av labradorsten. *Tidsskr. Kemi Bergv.*, 1924, **4**, 116–117.—*Tek. Ukebl.* 229.

Aluminiumoksyd av blaalere. *Tidsskr. Kemi Bergv.*, 1924, **4**, 157–160.—*Tek. Ukebl.* **71**, 380–381.

Om fordelingslovene for de sjeldne jordarter. Foredrag 22. febr. 1924. *Overs. Vidensk. møter i* 1924. Oslo 1925, s. 5.

1925

Geochemische Verteilungsgesetze der Elemente. 4. Zur Krystallstruktur der Oxyde der seltenen Erdmetalle. Von V. M. G., F. Ulrich und T. Barth. Mit 2 Textfig. Utg. for Fridtjof Nansens fond. 24 s. *Skrifter utg. av det Norske Videnskaps-akademi i Oslo. I. Mat.-naturv. klasse 1925.* No. 5.

Geochemische Verteilungsgesetze der Elemente. 5. Isomorphie und Polymorphie der Sesquioxyde. Die Lanthaniden-Kontraktion und ihre Konsequenzen. Von V. M. G., T. Barth und G. Lunde. Mit 6 Textfig. Utgit for Fridtjof Nansens fond. 59 + upg. s. *Skrifter utg. av det Norske Videnskaps-akademi i Oslo. I. Mat.-naturv. klasse 1925.* No. 7.

Om en forekomst av fossilførende underkambrium ved Usteberget syd for Ustaoset. Foredrag 23. jan. 1925. *Det. Norske Videnskaps-akademi i Oslo. Årbok 1925.* Oslo 1926, 3–4.

Om isomorfi og polymorfi. Ref. av foredrag 22. mai 1925. *Det Norske Videnskaps-akademi i Oslo. Årbok 1925.* Oslo 1926, 15–16.

(Svar til professor W. C. Brøgger i anledning av dennes meddelelse om at V. M. G. har avslått en kallelse til høiskolen i Charlottenburg.) *Det Norske Videnskaps-akademi i Oslo. Årbok 1925.* Oslo 1926, 31; jfr. 28–30 av W. C. Brøgger.

Forskningsarbeide over industrielle problemer. *Norges industri* 1925, 177–190.

Über fossilführende untercambrische Basalablagerungen bei Ustaoset. *Fennia,* Helsingfors **45**, 1–11.

1926

Geochemische Verteilungsgesetze der Elemente. 6. Über die Kristalstrukturen vom Rutiltypus, mit Bemerkungen zur Geochemie zweiwertiger und vierwertiger Elemente. (Von) V. M. G., nach Untersuchungen gemeinsam mit T. Barth, P. Holmsen, G. Lunde, W. Zachariasen. Utg. for Fridtjof Nansens fond. 21 + up. s. *Skrifter utg. av det Norske Videnskaps-akademi i Oslo 1926. I. Mat.-naturv. klasse.* No. 1.

Geochemische Verteilungsgesetze der Elemente. 7. Die Gesetze der Krystallochemie. (Von) V. M. G., nach Untersuchungen gemeinsam mit T. Barth, G. Lunde, W. Zachariasen. Utg. for Fridtjof Nansens fond. 117 + upg. s. *Skrifter utg. av det Norske Videnskaps-akademi i Oslo 1926. I. Mat.-naturv. klasse.* No. 2.

Om verdens kjemiske sammensetning. Ref. av foredrag 22. jan. 1926. *Det. Norske Videnskaps-akademi i Oslo. Årbok* 1926. Oslo 1927, 3–4.

Mindetale over professor Th. H. Hiortdahl holdt i den mat.-naturv. klasses møte 26. mai 1926. *Det Norske Videnskaps-akademi i Oslo. Årbok 1926.* Oslo 1927, 65–69.

Om sammensetningen av norsk lere, efter undersøkelser av foredragsholderen, J. Grenness, H. Hougen, H. Johnson og O. Rove. Foredrag 21. mai 1926. *Det Norske Videnskaps-akademi i Oslo. Årbok 1926.* Oslo 1927, 23–24.

Naturherligheterne i den norske fjeldgrund og i jordbunden. *Norge Tidsskr. om vårt land,* 1926, **2**, 354–360.

Zur Formel des Helvins. *Zentbl. Miner.*, 1926, **17**, Abt. A. 148–149.

Metasomatic processes in fissure veins. *Econ. Geol.*, 1926, **21**, 91–94.

Probleme und Methhoden der Geochemie. *G. Beitr. Geophys.*, 1926, **15**, 38–50.

Über das krystallochemische und geochemische Verhalten des Germaniums. *Naturwissenschaften, Berlin,* 1926, **14**, 295–297.

Die Gesetze der Krystallochemie. *Naturwissenschaften, Berlin,* 1926, **14**, 477–485.

Undersøkelser over lersedimenter. Foredrag. *Nord. JordbrForsk.*, 1926, Kbh. 434–445.

Bemerkungen zu der Arbeit von E. Günther und Wilcke: Beiträge zur Röntgenspektroskopie. II. *Z. phys. Chem.*, 1926, **122**, 250–253.

1927

Geochemische Verteilungsgesetze der Elemente. 8. Untersuchungen über Bau und Eigenschaften von Krystallen. Utg. for Fridtjof Nansens fond. 1927. 156 s. [S. 140–156: Anhang: Krystallstruktur-Bestimmungen aus dem Mineralogischen Institut der Universität Oslo (1923–1926).] *Skrifter utg. av det Norske Videnskaps-akademi i Oslo 1926. I. Mat.-naturv. klasse.* No. 8.

Om aluminiumhydroxyd som bestanddel av norsk lere. Ref. av foredrag 8. april 1927. *Det Norske Videnskaps-akademi i Oslo. Årbok 1927.* Oslo 1928, 12–13.

Relasjonene mellem stoffenes sammensetning og deres krystallform. Ref. av foredrag 20. mai 1927. *Det Norske Videnskaps-akademi i Oslo.* Årbok 1927. Olso 1928, s. 22.

Krystallbau und chemische Zusammensetzung. *Ber. dt. chem. Ges.*, 1926, **60**, 1263–1296.

A physico-chemical theory of metasomatism. *Econ. Geol.*, 1926, **22**, 521–523.

Über das Wählersche Cyanstickstofftitan. *Nachr. Ges. Wiss. Göttingen, Mat. phys. Kl.*, 390–393.

Konstruktion von Kirstallen. *Z. tech. Phys.*, 1928, **8**, 251–264.

1928

Om dannelse av laterit som forvitringsprodukt av norsk labradorsten. *H. Sørlie* 1878—27. okt. 1928. 21-24.

Beriktigelse (til Tom Barths oversikt over relasjonene mellem krystallstruktur og kjemisk sammensetning). *Naturen.* Bergen **52,** 319-320; jfr. 285-286 av T. Barth.

Darstellung und Eigenschaften des krystallisierten Bornitrides. *Norsk geol. Tidsskr.*, 1928, **9,** 258-265.

Über die Raumerfüllung der Atome (Ionen) in Kristallen und über das Wesen der Lithosphäre. *Neues Jb. Miner. Beilage.* Abt. A., 1928, **57,** 1119-1130.

Der Kristallbau und die Arten der chemischen Bindung. *Z. Elektrochem.* Lpz. 453-463.

Bermerkungen zu dem Aufsatze von Vegard: Die Röntgenstrahlen im Dienste der Erforschung der Materie. *Z. Kristallogr.*, 1928, **68,** 467-469.

Über Atomabstände in Metallen. *Z. phys. Chem.* Lpz., 1928, **133,** 397-419.

1929

Krystal-Kemi. *Beretning om det 18. skand. naturforskermøde København 1929.* Kbh. 1929, 114-118.

Krystal-Kemi. *Svenska fysikersamf. publ. Kosmos 7.*

Die Krystallstruktur des Rheniums. *Naturwissenschaften,* 1929, **17,** 134-135.

Geochemische Verteilungsgesetze der Elemente. *Phys. Z.*, 1929, **30,** 519-520.

The distribution of the chemical elements. *Proc. R. Instn Gt Br.*, 1929, **26,** 73-86, tabel.

Die Naturgeschichte der Eisenfamilie. *Stahl und Eisen,* 1929, **49,** 601-612.

Crystal structure and chemical constitution. *Trans. Faraday Soc.*, 1929, **25,** 253-283.

Die Röntgenstrahlen im Dienste der Erforschung der Materie. Erwiderung. *Z. Kristallogr.*, 1929, **69,** 213-214.

Die Gitterkonstante des Bariumtellurides. *Z. Kristallogr.*, 1929, **69,** 411-414.

Kristallstruktur, Gitterkonstanten und Dichte des Rheniums. *Z. phys. Chem. Abt. B.*, 1929, **2,** 244-252.

Bemerkung zu einer Abhandlung von A. M. Berkenheim, Über neue Gesetzmässigkeiten in den Reihen der elementaren Ionenradien. *Z. phys. Chem. Abt. A.*, 1929, **141,** 451-452.

1930

[Selvbiografi.] *Studentene fra* 1905. Biografiske oplysninger samlet til 25-årsjubileet 1930. Oslo 1930, 115.

Elemente und Minerale pegmatitischer Gesteine. Vorgelegt 9. Mai 1930. *Nachr Ges. d. Wiss. Göttingen. Math.-phys. Kl.*, 1930, 370-378.

Über das Vorkommen des Germaniums in Steinkohlen und Steinkohlen Produkten. *Nachr. Ges. Wiss. Göttingen. Math-phys. Kl.*, 1930, 398-401.

Geochemische Verteilungsgesetze und kosmishe Haüfigkeit der Elemente. *Naturwissenschaften,* 1930, **48,** 999-1013.

Über das Vorkommen des Germaniums im Meteoriten von Cranbourne. *Z. phys. Chem. Abt. A.*, 1930, **146,** 404-405.

1931

Über die kristallchemischen Beziehungen zwischen Gallium und Aluminium, Germanium und Silicium. *Norsk geol. Tidsskr.*, 1931, **12,** 247-264.

Kristallchemie und Röntgenforschung. *Ergebnisse techn. Röntgenkunde,* 1931, **2,** 152-182.

Kristallchemie. *Fortschr. Miner. Kristallogr. Petrogr.*, 1931, **15,** 73-146.

Der Kreislauf der Metalle in der Natur. *Metallwirtschaft,* Berlin, 1931, **10,** 265-272.

Zur Geochemie des Galliums. Von V. M. G. und Cl. Peters. Vorgelegt, 20. Febr. 1931. *Nachr. Ges. Wiss. Göttingen. Math.-phys. Kl.*, 1931, 165-183.

Zur Kristallchemie des Germaniums. Vorgelegt 22. Mai 1931. *Nachr. Ges. Wiss. Göttingen. Math.-phys. Kl.*, 1931, 184-190.

Zur Geochemie des Scandiums. Von V. M. G. und Cl. Peters. Vorgelegt 4. Dezember 1931. *Nachr. Ges. Wiss. Göttingen. Math. phys. Kl.*, 257-279. Friedrich Becke. *Nachr. Ges. Wiss Göttingen. Geschäftliche Mitteilungen.* 1931-1932, 70-73.

Kristallchemie. *UnterrBl. Math. Naturw.*, 1931, **37,** 273-279.

1932

Geochemische Betrachtungen. *Fortschr. Miner. Kristallogr. Petrogr.*, 1932, **17,** 415-417.

Isomorphie von Boraten und Karbonaten. Von V. M. G. und H. Hauptmann. Vorgelegt 12, Februar 1932. *Nachr. Ges. Wiss. Göttingen. Math.-phys. Kl.*, 1932, 53-72.

Zur Geochemie des Berylliums. Von V. M. G. und Cl. Peters. Vorgelegt 27. Mai 1932. *Nachr. Ges. Wiss. Göttingen. Math.-phys. Kl* 1932, 360-376.

Zur Geochemie der Edelmetalle. Von V. M. G. und Cl. Peters. Vorgelegt 27. Mai 1932. *Nachr. Ges. Wiss. Göttingen. Math.-phys. Kl* 1932, 377-401.

Zur Geochemie des Bors. Von V. M. G. und Cl. Peters. Vorgelegt 22. Juli 1932. *Nachr. Ges. Wiss. Göttingen. Math. phys. Kl* 1932, 402-407.

Zur Geochemie des Bors. II. Von V. M. G. und Cl. Peters. Vorgelegt 28. Okt. 1932. *Nachr. Ges. Wiss. Göttingen. Math.-phys. Kl* 1932, 528-545.

Otto Mügge. *Nachr. Ges. Wiss. Göttingen, Geschäftliche Mitteilungen 1932/33,* 42-46.

Neue Wege und Gesichtspunkte bei der Synthese von Mineralien und Gesteine. *Naturwissenschaften, Berlin,* 1932, **20,** 337-340.

Geochemische Leit-Elemente. *Naturwissenschaften, Berlin,* 1932, **20,** 947-948.

Die Rutilmodifikation des Germaniumdioxyds. *Z. phys. Chem. Lpz.*, 1932, **17,** 172-176.

1933

Grundlagen der quantitativen Geochemie. *Fortschr. Miner.*, 1933, **17,** 112–156.
Zur Geochemie des Germaniums. V. M. G. und Cl. Peters. Vorgelegt 37. Januar 1933. *Nachr. Ges. Wiss. Göttingen. Math.-phys. Kl.*, 1933, 141–166.
Zur Geochemie der Alkalimetalle. Von V. M. G., H. Berman, H. Hauptmann, Cl. Peters. Vorgelegt 24. März 1933. *Nachr. Ges. Wiss. Göttingen. Math.-phys. Kl.* 1933, 235–244.
Zur Geochemie des Selens. Von V. M. G. und O. Hefter. Vorgelegt 24. März 1933. *Nachr. Ges. Wiss. Göttingen. Math.-phys. Kl.*, 1933, 245–252.
Zur Kenntnis der Troilit-Knollen der Meteoriten. Ein Beitrag zur Geochemie von Chrom, Nickel und Zinn. Von V. M. G. und Cl. Peters. Vorgelegt 19. Mai 1933. *Nachr. Ges. Wiss. Göttingen. Math.-phys. Kl.* 1933, 278–287.
Über die Anreicherung seltener Elemente in Steinkohlen. Von V. M. G. und Cl. Peters. Vorgelegt 28. Juli 1933. *Nachr. Ges. Wiss. Göttingen. Math.-phys. Kl.* 1933, 371–386.
Über die Berücksichtigung »seltener« Elemente bei Gesteins-Analysen. Von V. M. G., H. Hauptmann, Cl. Peters. *Naturwissenschaften,* 1933, **21,** 363–365.

1934

Kristallchemie. *Handwörterbuch Naturw.* 2. Aufl. 5, 1128–1154.
Drei Vorträge über Geochemie. *Geol. För. Stockh. Förh.*, 1934, **56,** 385–427.
Zur Geochemie des Arsens. Von V. M. G. und Cl. Peters. Vorgelegt 26. jan. 1934. *Nach. Ges. Wiss. Göttingen. Math.-Physikal Kl. N. F. Fachgruppe IV.* 1 (1934–40). Göttingen 1941, 11–22.
Zur Geochemie der Alkalimetalle. II. Von V. M. G., H. Bauer und H. Witte, Vorgelegt 27. Juli 1934. *Nachr. Wiss. Göttingen. Math.-phys. Kl. N. F. Fachgruppe* IV. 1 (1934–40). Göttingen 1941, 39–55.
W. Biltz: Raumchemie der festen Stoffe. (Anm.) *N. F. Fachgruppe* IV. 1 (1934–40), 123–142.

1935

Problemstillinger, arbeidsmetoder og resultater ved de geokjemiske undersøkelser i Göttingen 1929–1935. Foredrag 4. okt. 1935. *Det norske videnskaps-akademi i Oslo. Årbok 1935.* Oslo 1936, s. 23.
Grundlagen der quantitativen Geochemie II. *Fortschr. Miner.*, 1935, **19,** 183–216.
Rare elements in coal ashes. *Ind. Engng Chem.*, 1935, **27,** 1100–1102.
Zur Geochemie des Selens. II. Von V. M. G. und L. W. Strock. Vorgelegt 22. Februar 1935. *Nachr. Ges. Wiss. Göttingen. Math.-phys. Kl. N. F. Fachgruppe IV.* 1 (1934–40), 123–142.

1936

Geokjemiske forskningsarbeider i Göttingen, metallenes fordeling i naturen. *Tek. ukebl.*, 124–128.

1937

Geochemie und periodisches System der chemischen Elemente. *Travaux du congrès jubilaire Mendeléev II. Académie des Sciences de l'URSS.* 387–396.
The principles of distribution of chemical elements in minerals and rocks. *J. Chem. Soc., London,* 655–673.

1938

Geochemische Verteilungsgesetze der Element. 9. Die Mengenverhältnisse der Elemente und der Atom-Arten. Mit 6 Textfig. Utg. for Fridtjof Nansens fond 1938. 148 s. *Skrifter utg. av det Norske Videnskaps-akademi i Oslo 1937. I. Math-naturv. klasse.* No. 4.
De senere års utvikling av vårt kjennskap til geokjemien. Luftens, havets og jordens mengdevise opbygning av elementene. *Naturen,* 1938, **62,** 321–334, 353–362.
(Diskusjonsinnlegg til Olaf Holtedahls og Tom. Barths foredrag om Oppdalsbladet og sparagmittens metamorfose.) *Norsk geol. Tidsskr.*, 1938, **18,** 77–78.
Fordelingen av de kjemiske elementer i naturen. Utdrag av et foredrag. *Tidskr. Kjemi Bergv.*, 1938, **18,** 53–56.
Olivine and forsterite refractories in Europe. *Ind. Engng Chem.*, 1938, **30,** 32–34.

1939

Den kjemiske sammensetning av jorda og verda. *Universitetets radioforedrag. Jorda og livet på jorda gjennom tidene. Utg. av Norsk rikskringkasting,* 7–36.
Nye trekk i verdens råstofføkonomi. Foredrag 10. febr. 1939. *Det Norske Videnskaps-akademi i Oslo. Årbok 1939.* Oslo 1940, 3.
Spektralanalysens anvendelse i mineralogi, geokjemi og metallurgi. Foredrag. *Tidsskr. Kjemi Bergv.*, 1939, **19,** 82–86.

1940

W. C. Brøgger som mineralog og petrograf. Minnetale holdt i Akademiets møte 8. mars 1940. *Ved W. C. Brøggers bortgang. Oslo 1940.* 15–29.—*Det Norske Videnskaps-akademi i Oslo. Årbok 1940, Oslo 1941,* 29–41.
(Innlegg i diskusjonen etter Anders Kvales foredrag: Petrografisk-tektoniske undersøkelser i fjellkjeden mellom Bergens-buene og Voss.) *Norsk geol. Tidsskr.*, 1940, **21,** 198.
Nye ildfaste materialer av norske råstoffer. Foredrag. *Teknisk ukeblad,* 1–7.
Professor dr. W. C. Brøgger, *Tidsskr. Kjemi Bergv.*, 1940, **20,** 21.

1941

Trekk av Sunnhordlands geologi. Innlegg etter Niels-Henrik Kolderups foredrag. *Norsk geol. Tidsskr.*, 1941, **21,** 185–186.

Trekk av Narvik-Ofoten-traktens geologi. Innlegg etter Th. Vogts foredrag. *Norsk geol. Tidsskr.*, 1941, **21,** 202–210.

1942

Om super-uraner, grunnstoffer med større kjerneladning enn 92. *Fra fysikkens verden. Oslo.* **3** (1941/42), 179–190.

1943

The occurrence of rare elements in coal ashes. *B. C. V. R. A. Research Staff.* 5 s.

Om friksjonsglass (pseudotachylitt) i fjellkjeden. *Geol. För Stockh. Förh.*, 1943, **65,** 83–84.

Oksydasjon og reduksjon i geokjemien. *Geol. För. Stockh. Förh.*, 1945, **65,** 84–85.

Geokemiska undersökningar. *Geol. För. Stockh. Förh.*, 1943, **65,** 310–312.

1944

Geokjemiske resultat om atomenes mengdeforhold og stabilitet i universet. *Norsk geol. Tidsskr.*, 1944, **22,** 210–211.

Crystal chemistry and geochemistry. *Chemical Products,* March–Apr., 1–6.

Presentation of the Wollaston Medal. *Proc. Geol. Soc. London.* May 1944.

1945

The effect of olivine on the lungs of rats. (Sammen med:) E. J. King, Nancy Rogers, Margaret Gilchrist, G. Nagelschmidt. *J. Path. Bact.*, 1945, **57,** 488–491.

The geochemical background of minor-element distribution. *Soil Sci.* **60.**

1946

Anvendelse av olivinstein i støperier. *Teknisk Ukeblad.* No. 50.

Ionic radius of divalent copper. *Nature,* 1946, **157,** 192–193.

Krystallografi. Forelesninger til den mat.-naturv. embedseksamens almindelige avdeling. Kra. 1919–20. *Geol. mus. bibl.* IV. 4°.

Forelesninger i mineralogi (Hektografert.) *Geol. mus. bibl. Arkivrommet.*

Krystallografi. (Maskinskrevne forelesninger.) Oslo 1920–1931. *Geol. Mus. bibl.* IV. 4°.

Forelesninger i mineralogi. 1. oppl. Göttingen 1930. 2. oppl. Oslo 1936. 3. oppl. Oslo 1940–41. (Mangfoldiggjort som ms.) *Geol. mus. bibl. IV.* 4°.

Forelesninger over petrografi. 2. utg. Oslo 1940–41. *Geol. mus. bibl. IV.* 4°.

Formann i den av Industriforsyningsdepartementet 16. nov. 1917 nedsatte konsultative komité til bistand for dep. tel. til opplysning om landets resurser av malmer og bergarter samt deres nyttiggjørelse på de områder, hvor dette naturlig trengtes. Se for øvrig V. Haffner: Innstillinger og betenkninger 1814–1925. Oslo 1925, 1277.

Medlem av den av Industriforsyningsdepartementet 15. mars 1918 nedsatte komité til å utrede spørsmålet om opprettelsen av et forskningsinstitutt for ildfaste materialer. Om de av komitéen forskjellige uttalelser henvises til V. Haffner: Innstillingen og betenkninger 1814–1924. Oslo 1925 No. 1297.

Medlem av den av Kirkedepartementet 4. okt. 1920 nedsatte komité for å undersøke en mulig endring av Norges geologiske undersøkelses organisasjon. Komitéens innstilling er i ms. og dat. 21. jan. 1921 og utgjør 18 + 2 s. 4°.

Medlem av den under 8. mai 1922 av Kirkedepartementet nedsatte komité for å oppnå et organisk samarbeide mellem de videnskapelige forskningsfond.

POSTHUMOUS

Geochemistry, edited by Alex Muir. Clarendon Press, Oxford, 1954.

Appendix C

MEDICAL RECORD OF V. M. GOLDSCHMIDT

Dr. Leiv Gjessing, who studied mineralogy with Goldschmidt before taking a medical degree, has kindly contributed the following account of his medical history.

His father had diabetes. Victor Moritz Goldschmidt had pneumonia when he was four and eight years old, and was operated for appendicitis in 1909, for peritonitis in 1919, and for hernia twice in 1924. In 1917–18 he had rheumatic fever. In 1928 he spent a month at the spa of Bad Ems "recovering from a catarrh of the throat, which I have got from the combined action of HF fumes and an influenza". In 1933 he began to be plagued by shortness of breath during physical work. X-ray examination showed distinct enlargement of the heart. Since then he was occasionally bothered by subjective heart symptoms as well as slight swelling of the feet.

In 1937 he had angina with high fever over several days. In the following two weeks he was very weak and tired. He was therefore admitted to Lovisenberg Hospital in Oslo, where he stayed 3–4 weeks. Albuminuria was detected and it was unchanged when he was discharged. Some months later he discovered for the first time marked swelling of the feet. Since then he was troubled with headaches, which however did not prevent him from working.

In spite of a document, signed by Hitler and Göring, giving him permission to leave Germany and travel to Norway in 1935, he was arrested in October 1942 and put into a concentration camp for Jews. After a few days he became very ill. He was admitted to the local hospital in Tønsberg on October 31, 1942, because of severe pains in the back, headache, nausea, and disturbances of vision. Medical examination showed: swelling of the feet, albuminuria (Esbach 1 pro mille), hypertension (210/130), hypertrophy of the heart (but the heart sounds were normal and the pulse was regular), as well as polycythemia (sedimentation rate 1 mm, hemoglobin 135%, red cells 9,580,000). ECG: considerable left ventricular hypertrophy, PQ 0.19, T1 neg., T2 biophasic, T3 pos., T4 neg., P–waves broad and split: cardiomyopathy. His symptoms regressed gradually and within 5–6 days he felt much better. He left the hospital on November 8, 1942 for Oslo, as he was told he was released from the concentration camp.

Later he was transported to Sweden, where in February 1943 he spent four weeks in Ersta Hospital, Stockholm, for the treatment of internal bleeding and heart disease. In the spring of 1943 he was brought by air to England and Scotland.

In Aberdeen, in the fall of 1943, he spent six weeks in the Royal Infirmary for polycythemia, which was successfully treated with deep X-rays. In July 1944 he spent ten days in an Edinburgh clinic for treatment of heart disease. In December 1944 he suffered a near–fatal heart attack while he was a guest in Dr. W. G. Ogg's house in Harpenden, resulting in a lengthy hospital stay. On the recommendation of his doctors, on leaving the hospital he entered Miss Debenham's Nursing Home in Harpenden, where emergency treatment was available in case of another heart attack. He remained in the nursing home, apart from occasional visits to London, until his return to Norway in June 1946. His experiences during the war years had led to a serious deterioration in his health, and when he returned to Norway he was very weak both physically and mentally.

One night shortly after his return to Norway he had a severe heart attack with pulmonary edema, which was successfully treated at the National Hospital in Oslo. Later, in the fall, he received X-ray treatment of a melanoma on his leg. In the spring of 1947 he had to be operated. When he came home from the hospital on March 20 he complained of intense pain in his head and died suddenly.

In summary the diagnoses were:

1892: Pneumonia
1896: Pneumonia
1909: Appendicitis
1917–18: Rheumatic fever, causing his heart disease
1919: Peritonitis
1924: Hernia
1937: Angina with high fever, causing renal symptoms
1942: Hypertension, heart disease, and polycythemia
1943: Internal bleeding and heart disease
1943: Polycythemia, successfully treated
1944: Heart attack
1944: Heart attack, near fatal
1946: Heart attack with pulmonary edema
1947: Malignant melanoma and pulmonary fibrosis

Appendix D

Scientific Honors

Fridtjof Nansen Prize, Norwegian Academy of Sciences, 1912
Knight of the Royal Norwegian Order of St. Olav, 1929

Honorary doctorates

University of Freiburg (Breisgau), 1932
University of Utrecht, 1936
University of Aberdeen, 1944

Memberships in Academies of Sciences

R. Norwegian Academy of Sciences, 1914 (member)
R. Swedish Academy of Sciences, 1929 (foreign member)
R. Danish Academy of Sciences, 1928 (foreign member)
R. Swedish Academy of Engineering Science, 1921 (foreign member)
Regia Accademia dei Lincei, Rome, 1929 (foreign member)
Gesellschaft der Wissenschaften, Göttingen, 1929 (member)
Akademie der Wissenschaften, Vienna, 1930 (foreign correspondent)
U.S.S.R. Academy of Science, 1926 (foreign correspondent)
Royal Society, London, 1943 (foreign member)
Royal Society of Science, Uppsala, 1940 (foreign member)
Royal Society of Physiography, Lund, 1924 (foreign member)
Academy of Sciences, Halle, 1927 (foreign member)

Honorary memberships in learned societies

Geological Society of Stockholm
Geological Society of Finland
Geological Society of London
Geological Society of Edinburgh
Mineralogical Society of Great Britain
Geological Society of America
Mineralogical Society of the U.S.S.R.
Chemical Society (of London)

Appendix E

THE CHEMICAL ELEMENTS AND THEIR SYMBOLS

	Symbol	Atomic number		Symbol	Atomic number
Aluminum	Al	13	Holmium	Ho	67
Antimony	Sb	51	Hydrogen	H	1
Argon	Ar	18	Indium	In	49
Arsenic	As	33	Iodine	I	53
Barium	Ba	56	Iridium	Ir	77
Beryllium	Be	4	Iron	Fe	26
Bismuth	Bi	83	Krypton	Kr	36
Boron	B	5	Lanthanum	La	57
Bromine	Br	35	Lead	Pb	82
Cadmium	Cd	48	Lithium	Li	3
Calcium	Ca	20	Lutetium	Lu	71
Carbon	C	6	Magnesium	Mg	12
Cerium	Ce	58	Manganese	Mn	25
Cesium	Cs	55	Mercury	Hg	80
Chlorine	Cl	17	Molybdenum	Mo	42
Chromium	Cr	24	Neodymium	Nd	60
Cobalt	Co	27	Neon	Ne	10
Copper	Cu	29	Nickel	Ni	28
Dysprosium	Dy	66	Niobium	Nb	41
Erbium	Er	68	Nitrogen	N	7
Europium	Eu	63	Osmium	Os	76
Fluorine	F	9	Oxygen	O	8
Gadolinium	Gd	64	Palladium	Pd	46
Gallium	Ga	31	Phosphorus	P	15
Germanium	Ge	32	Platinum	Pt	78
Gold	Au	79	Potassium	K	19
Hafnium	Hf	72	Praseodymium	Pr	59
Helium	He	2	Radium	Ra	88

	Symbol	Atomic number		Symbol	Atomic number
Radon	Rn	86	Terbium	Tb	65
Rhenium	Re	75	Thallium	Tl	81
Rhodium	Rh	45	Thorium	Th	90
Rubidium	Rb	37	Thulium	Tm	69
Ruthenium	Ru	44	Tin	Sn	50
Samarium	Sm	62	Titanium	Ti	22
Scandium	Sc	21	Tungsten	W	74
Selenium	Se	34	Uranium	U	92
Silicon	Si	14	Vanadium	V	23
Silver	Ag	47	Xenon	Xe	54
Sodium	Na	11	Ytterbium	Yb	70
Strontium	Sr	38	Yttrium	Y	39
Sulfur	S	16	Zinc	Zn	30
Tantalum	Ta	73	Zirconium	Zr	40
Tellurium	Te	52			

Appendix F

THE GEOLOGICAL TIME SCALE (M = MILLIONS OF YEARS)

Eras	Approximate age (in years)	Subdivisions	Approximate duration (in years)
CENOZOIC	10,000	RECENT	10,000
	2 M	PLEISTOCENE	2 M
	5 M	PLIOCENE	3 M
	25 M	MIOCENE	20 M
	35 M	OLIGOCENE	10 M
	67 M	EOCENE	32 M
MESOZOIC	140 M	CRETACEOUS	73 M
	210 M	JURASSIC	70 M
	250 M	TRIASSIC	40 M
PALEOZOIC	290 M	PERMIAN	40 M
	360 M	CARBONIFEROUS	70 M
	410 M	DEVONIAN	50 M
	440 M	SILURIAN	30 M
	500 M	ORDOVICIAN	60 M
	590 M	CAMBRIAN	90 M
PRECAMBRIAN		Worldwide subdivisions not well established	3200 M+
		Oldest rock dated 3800 M	
		Age of the earth 4500 M	

Appendix G

GLOSSARY

ACTINIDE: one of the elements following actinium (atomic number 89) in the Periodic Table

ALKALINE EARTHS: the elements in Group 2A of the Periodic Table (beryllium, magnesium, calcium, strontium, barium)

AMPHIBOLE: one of a group of common minerals with the general formula $A_{2-3}B_5(Si,Al)_8O_{22}(OH)_2$, in which A = Ca, Na, K; B = Mg, Fe, Al

ANATEXIS: the melting of a preexisting rock by natural processes

ANION: a negatively charged ion, produced when a neutral atom accepts one or more additional electrons

ANTICATHODE: the target in an electron tube for the production of X-rays

ATOMIC NUMBER (Z OR AN): the number of protons in the nucleus of an atom, or the total number of electrons in the neutral atom

BIOSPHERE: the sum total of living matter

BIOTITE: brown or black mica, with the formula $K(Mg,Fe)_3(AlSi_3O_{10})(OH)_2$

CATION: a positively charged ion, produced by the loss of one or more electrons from a neutral atom

CHROMITE: the principal mineral of chromium, with the formula $(Mg,Fe)(Cr,Al)_2O_4$

DIFFERENTIATION: the physical segregation of one type of material from another; a homogeneous or intimately mixed body becomes segregated into two or more phases of different composition

DOCENT: in Europe, a qualified university teacher, approximately equivalent to associate or assistant professor in North America

DOLOMITE: calcium–magnesium carbonate, $CaMg(CO_3)_2$, or a rock consisting essentially of this mineral

ELEMENT: a substance containing a single type of atom (all atoms having the same atomic number Z)

EQUILIBRIUM: the state of a chemical system in which the phases do not undergo any change with the passage of time

FELDSPAR: a group of minerals, including the alkali feldspars, with compositions ranging from

$KAlSi_3O_8$ to $NaAlSi_3O_8$, and the plagioclase feldspars, with compositions ranging from $NaAlSi_3O_8$ to $CaAl_2Si_2O_8$

GALENA, GALENITE: lead sulfide, PbS; the principal ore of lead

GEOBAROMETRY: the measurement of the pressure of formation of a rock

GEOTHERMOMETRY: the measurement of the temperature of formation of a rock

GNEISS: a metamorphic rock, in which bands or lenses of granular minerals alternate with bands or lenses of flaky or prismatic minerals

GONIOMETER: an instrument for measuring the angles between crystal faces

HALF-LIFE: the time necessary for half the atoms of a radioactive substance to disintegrate

HYDROLYSIS: the reaction between water and a chemical compound, often resulting in the precipitation of a metal hydroxide

HYDROSPHERE: the discontinuous shell of water—fresh, salt, and solid—at the surface of the Earth

HYDROTHERMAL: refers to hot aqueous solutions circulating in the Earth's crust, or to veins and mineral deposits crystallized from such solutions

ION: an atom with a net electrical charge, due to the acquisition or loss of electrons

IONIC POTENTIAL: the ratio of an ion's charge to its radius

ISOTOPE: atoms of an element whose nuclei differ in the number of contained neutrons

KILOBAR (kbar): a unit of pressure equal to 1000 bars (one bar is approximately one atmosphere); it has been superseded in the SI system by the pascal (Pa), 1 bar = 10^5 Pa

LANTHANIDE: one of the fourteen elements following lanthanum in the Periodic Table

MAGMA: a rock melt within the Earth's crust; lava is magma poured out of volcanoes

METAMORPHISM: the sum of the processes working within the Earth's crust to cause the recrystallization of rocks. Contact metamorphism is produced by the heat and pressure of intruding magma; regional metamorphism covers extensive areas, usually in association with mountain-building processes

MIGMATITE: a composite rock consisting of igneous and metamorphic material; the term was introduced by J. J. Sederholm in 1907 to describe such rocks in the Precambrian of Finland

NEUTRON: an uncharged nuclear particle

OROGENY: the process of the formation of mountain chains: the Caledonian orogeny, dated as late Silurian, extended from Scotland (Caledonia) and Ireland northeastward through Scandinavia

PARAGENESIS: a characteristic association or occurrence of minerals

PEGMATITE: an extremely coarse-grained igneous rock, usually found as irregular dikes, lenses, or veins

PETROLOGY: the scientific study of rocks

PHOSPHORESCENCE: the emission of light by a substance after irradiation

PLAGIOCLASE: see feldspar

POLYCYTHEMIA: an excess of red cells in the blood

POLYMORPH: one of alternative crystal structures a substance can adopt

PROTON: a positively charged nuclear particle

PYROLUMINESCENCE: the emission of light by a substance when heated

PYROXENE: one of a group of minerals with the general formula $R_2Si_2O_6$; in the common pyroxenes R = Ca, Fe, Mg and some of the Si may be replaced by Al

PYROXENITE: a rock consisting largely of pyroxene

REFRACTORY: having a high melting point

SCHIST: a metamorphic rock that can be split into thin slabs, due to the parallelism of flaky or prismatic minerals such as the micas or amphiboles

SILICOSIS: a lung disease caused by the inhalation of fine-grained quartz

SPECTROGRAPH: an instrument designed to photograph a spectrum, the collection of individual wavelengths of light

SYENITE: an igneous rock consisting largely of alkali feldspar with minor amounts of biotite and/or amphibole; it may contain a little quartz, but if more than 10% is present the rock grades into granite

TRIBOLUMINESCENCE: the emission of light when a substance is crushed

UNIT CELL: the simplest polyhedron, containing one or more formula units, which by repetition forms a crystal structure

VALENCY: the number of chemical bonds that an element can utilize in forming a compound

XENOLITH: a foreign inclusion in an igneous rock

Index